Games as Models of Social Phenomena

An individual decision with social consequences
(see Chapter 7)

"Now, surely, the mighty Con Ed isn't going to topple if I turn on one little itsy-bitsy lamp."

Games as Models of Social Phenomena

Henry Hamburger
University of California, Irvine

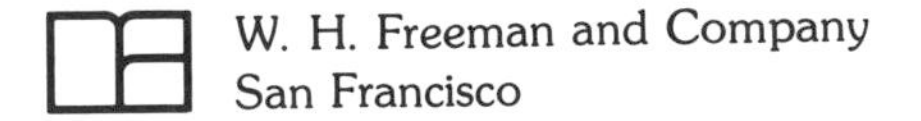
W. H. Freeman and Company
San Francisco

Sponsoring Editor: Richard J. Lamb
Project Editor: Pearl C. Vapnek
Manuscript Editor: Howard Beckman
Designer: Marjorie Spiegelman
Production Coordinator: Linda Jupiter
Illustration Coordinator: Cheryl Nufer
Compositor: Bi-Comp, Inc.
Printer and Binder: The Maple-Vail Book Manufacturing Group

Library of Congress Cataloging in Publication Data

Hamburger, Henry, 1940–
Games as models of social phenomena.

Bibliography: p.
Includes index.
1. Social sciences—Decision making. 2. Game theory. I. Title.
H61.H277 300′.1′5193 78-23267
ISBN 0-7167-1011-0
ISBN 0-7167-1010-2 pbk.

9 8 7 6 5 4 3 2 1

Contents

Preface

In 1963 the world's problems seemed a lot more significant to me than the graphs, matrices, and other abstractions that confronted me as a student in mathematics. And so I went out to do good in, and to try to understand, the world. Two years later I returned to graduate school, only to learn that the world's problems *were* graphs and matrices, or at least could be better analyzed in terms of them. Inspired by this understanding, I went off, in 1969, to share it with undergraduates. This book grows out of my attempts since then to do two things: to convince science and engineering students that "social science" is not a contradiction in terms, but a plausible and exciting objective, and to convince students in the social sciences and humanities that their conception of social and political life can be expanded and unified by an understanding of formal systems.

Game models attempt to make sense out of the complexity of human affairs by focusing on the interaction of decisions of either individuals or organizations whose behavior can potentially affect one another. We will see how informed self-interest, tempered by knowledge of the interests of others, may lead to mutual accommodation and group action. The games approach provides new insight into some of the traditional problems of inquiry of political science, psychology, and economics by placing them in an interdisciplinary framework. The nature of this unified view will become clear in the course of the book, but here we may take note of a few problems that can benefit from game theory.

In political science there has recently been a strong push to use notions of game theory in analyzing certain aspects of elections and legislative voting. Because votes are clear-cut decisions, they are a good place to start applying a theory of decision-making, and this is done in Chapter 8. Since records of voting are kept, data are often available to test the theory. Law enforcement and international relations are also amenable to analysis in terms of game theory, though the options facing the decision-makers are less clear-cut. These topics are treated in Chapter 4.

We will consider an economist's view of a pollution problem and how to solve it in Chapter 6. Various problems created by large-scale social organization, including energy distribution, population control, and highway congestion form the basis of Chapter 7. We will see that many such problems, and their potential solutions, can be better understood when formulated in the framework of game theory.

In addition to using game models to analyze real-world situations, it is also possible to create artificial games for use in social-psychological experiments, and this approach is surveyed in Chapter 9. Using remarkably simple games, it has been possible to evoke intense emotional involvement of human subjects and to gain some insight into how situations elicit behavior that may be described in terms of greed, cooperation, threat, vengeance, leadership, heroism, and martyrdom.

Various games that might ordinarily be regarded as political, economic, social, or psychological games are used throughout the book to illustrate game theory. Since game theory focuses on the decision structures of situations, the reader will find that the games used as examples are grouped according to the features of their decision structures they have in common, and not according to traditional disciplines. In Chapter 5, for example, we will regard convention delegates as involved in a game of timing rather than one of politics.

In writing this book, I have assumed that readers will be familiar with some high school mathematics, specifically, elementary algebra and the rudiments of probability. In addition, the reader must be able and willing to follow a complex line of thought, expressed partly in symbolic form. Much of the math is really nothing more than a translation of ordinary English into a language of symbols, which is more efficient once you are used to it. This should not be too surprising since you already know that ordinary language is too cumbersome for many ordinary tasks. Imagine, for example, trying to add up a grocery bill using words instead of numbers, or consider how many words would be needed to express the information given on a street map, where each line represents a street. Not only does the map require less space, but it is far easier to use. In any case, the symbols and mathematical manipulations in game theory are not intended as arbitrary hurdles, but as tools for clarification. It is my fond hope that conceptually difficult theoretical results will appear here as common sense to nontechnical readers and as thinly disguised proofs to the mathematician.

A word of explanation is needed about the use of pronouns. Some players turn out to be *he,* while others turn out to be *she*. For example, in the "Old Shell Game" described in Section 2.1, the reader will be partway through the example before discovering that Carnival-goer is female. This is an attempt to avoid any preconceived notions about sex roles and has no significance in game theory.

It is a pleasure to acknowledge some intellectual, spiritual, and practical contributors to this book. Anatol Rapoport, as mentor and model, helped me to see how analytic tools help make the world intelligible. Anyone who enjoys this book will, I believe, find his work compelling reading. Charlie Lave let me learn by doing as we co-taught the interdisciplinary model-building course at the University of California, Irvine. The relevance of my applications owes much to our conversations. John Boyd said he could "hear my voice" while reading an early draft of the book, and that got me through some dreary moments. Jerry Kaiwi caught a lot of relevant news stories I missed and pointed out their game-theoretic significance.

Of the hundreds of students to whom I have tried to convey the excitement and relevance of this material, I thank the many who have demanded the relevance and caught the excitement; among these, Mike Cane deserves special mention. The School of Social Sciences of the University of California at Irvine has been a congenial place to work, led by deans who have shared a commitment to the interdisciplinary environment in which this book took shape. Dick Lamb of W. H. Freeman and Company listened patiently while I told him why certain changes could not be made, and waited patiently while I made them. Howard Beckman provided invaluable editorial advice and assistance. The manuscript was typed and retyped with speed and good cheer by Kathy Alberti, Mary Bruce, Lillian White, Peggy Popovich, Donna Dill, and Ilsa Wilbert.

Games as Models of Social Phenomena

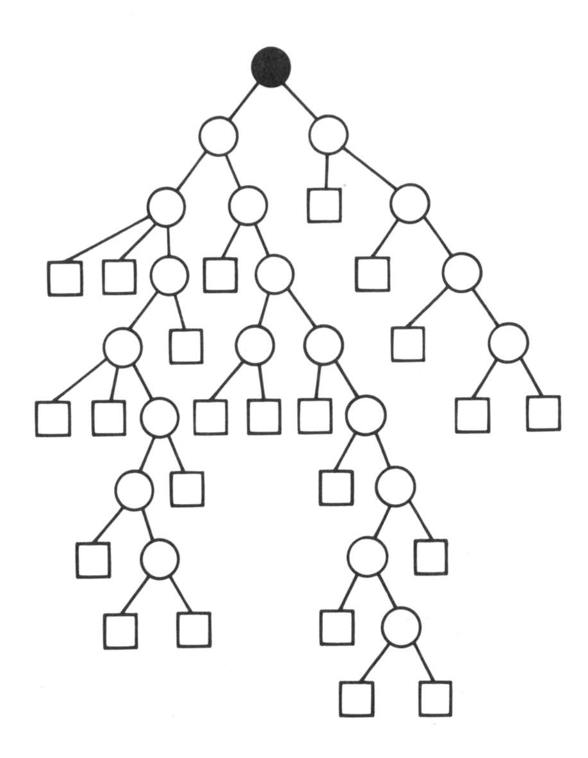

1 Orientation

Life is not a game, for life is real, while games are child's play. Yet this book is about something called *game theory* and how it can help us understand real-world situations and events. This is possible because the word "game" as used in game theory is not used in its ordinary sense. Just what that word *will* mean for us is discussed in the first section of this chapter. Certain uses of the word "game" are definitely not included in our study—specifically, simulation games and "games people play," which are discussed in Section 1.2. The final section of this chapter explains the interdisciplinary nature of the game-theory approach in the social sciences.

1.1 What Is a Game and Is Life One?

The word "game" brings to mind a variety of notions, including recreation, competition, sportsmanship, and strategic choice. For game theory, however, the key idea of a game is that the players make decisions that affect each other. By defining games in this way, we include not only tic-tac-toe and chess but also a wide variety of real-life situations from the domains of political science, economics, sociology, and social psychology.

Examples of real situations whose outcomes depend on the actions of several parties are easy to find. The outcome of the conflict in Vietnam depended not only on U.S. decisions to bomb, withdraw, "Vietnamize," and so on, but also on Vietcong decisions, not to mention the decisions of various other parties. The outcome of political action for civil rights depends not only on decisions made by activists, but also on those made by the rest of society, including police chiefs and senators. Unions and management, bidders at an auction, the Arab nations, Israel, and the Palestine Liberation Organization, the candidates for the 1980 Republican presidential nomination, the world's whalers overwhelming the world's whales on international waters, people waiting in line for concerts or gasoline—all of these, be they persons, organizations, or nations, are making decisions that have consequences not only for themselves but for others as well, and are in turn crucially affected by the decisions of the others.

Game theory is a unified approach to this broad array of political, social, and economic situations. A game-theoretic analysis requires answers to several key questions: Who has decisions to make? What are the different options available? What will be the results of the various possible combinations of choices? Which results are preferred by whom? To give some idea of what answers to these questions look like, let us consider a couple of real-life examples and try to answer the question of who the decision-makers are.

Several prison uprisings have taken place in recent years. Who is a decision-maker, or, in the language of game theory, who is a "player" in a prison uprising? Surely the prisoners play a role. If they take people hostage—usually some of the guards—then the hostages could be considered players. We could also include the rest of the guards, the chief warden of the prison, the governor of the state, reporters, citizens, the families of the hostages—all of these seem to be involved. Which of them to include in a game model depends partly on which aspect of the situation concerns us most and partly on how complex a model we can comprehend.

There are, however, two definite criteria that are used in picking out players for a game-theoretic analysis. First, because we are fundamentally concerned with decisions, we eliminate anyone who has no decisions to make. Second, because we are ultimately concerned with the results of decisions and the value of those results (payoffs) to various players, we want to disregard anyone who has no preference among the possible outcomes of the situation. In a prison uprising the hostages appear to have no alternative but to sit still and hope for the best. If we take this to be the case, then the hostages are not bona fide players, though of course they care more than anyone else about the outcome.

In contrast, it might happen that the person who makes the final, crucial decision is a judge or arbitrator who is supposed to be indifferent to the outcome, whose only responsibility is to "fairness" or "justice." This happened when the California Supreme Court decided that the Teamsters Union had colluded with fruit and vegetable growers against the United Farmworkers Union. The two unions and the growers were bona fide players, but the court was not, because (we assume) it would have received no payoff either way. A bona fide player is one who cares about what happens *and* can do something about it. The justices are not players because they are not supposed to care what happens; the prisoners' hostages care but cannot act.

Once the players have been identified, we can proceed to the other three points mentioned above: choices, results, and preferences. These aspects of the specification of a situation are placed into a game-theory framework in Chapter 2. Once a game is specified, we shall be concerned, in later chapters, with trying to "solve" it, that is, to come up with some idea of what options the players should select or probably will.

So far we have been emphasizing similarities between parlor games, such as tic-tac-toe and chess, and a broad range of real situations. We have done this by focusing on decision-making and interdependence of the participants. However, there are some very important differences between parlor games and the real situations that we are also calling games. For example, in a parlor game it is easy to see who the players are, while in a real situation it is more difficult, as the above examples show. Another way in which parlor games are more precisely defined than real-life situations is that in chess, for example, the rules tell you exactly what you are allowed to do, while the rules of life are not always clearly spelled out.

The most important way in which the typical parlor game differs from many real situations involves a key concept in game theory, the "zero-sum game." Such games are discussed at length in Chapter 3. For now, the main point is that in a zero-sum game one player can improve his outcome only at the expense of someone else. Parlor games have this property but most real situations do not. Thus, if you and I play chess, you can achieve a win only if I sustain a loss, but if two nations play nuclear confrontation, it is all too easy to imagine an outcome in which both of them "lose."

To take another example, the real game of buying and selling allows both players to win: if Agatha sells Bertram her petrified lotus leaf, she may be happy to have the money while he is delighted with the addition to his collection. Both of them have come out ahead according to their individual preferences or else they would not have agreed to the transaction. More generally, several individuals, organizations, or nations can do things that benefit or harm all of them, so that everyone may win or everyone may lose. Additional examples are legislative log-rolling, friends doing each other favors, corporate price-fixing, and international trade and tariffs.

The title of this section asks two things, what a game is and whether life is a game. We are now in a position to say yes, life is a game (or many games) provided "game" is defined as the interaction of decision-makers. This definition includes both real-life games and parlor games. What is the *purpose* of analyzing real situations as games? In other words, granted that life contains interactive-decision structures that are not zero-sum, what can the social scientist hope to learn by focusing on decision-making? Like other scientists, the social scientist wants to be able to predict what will happen in a given situation. By imagining himself in the role of first one player then another, by looking at the possible decisions facing each player, he may learn what choices would be optimal for each player. Assuming then that every player will actually see what is optimal for himself (and do it), the game-theorist can make a prediction about the overall outcome.

We need not restrict attention to the players. Equally fruitful is to look at a situation from the viewpoint of those persons who are in a position to change the rules or payoffs of the game. These "powers that be" in a nation, a school system, or a business office can create or modify the laws or rules and regulations that are in their domain of authority. By doing so, they alter the framework of decision or "rules of the game." Those who must make decisions within that framework must adapt their decision-making to such changes. Thus, Congress and the president enact corporate laws that affect the decisions of corporations, while school boards or individual teachers can create more (or less) competitive situations that may elicit less (or more) cooperation among students. The writing of treaties is an exercise in making up rules that will lead nations away from mutually destructive decisions in the future. The problem of creating sensible rules will be central to Sections 4.1, 4.2.1, 6.2, and 7.3.

Notice that a person or group may be the rule-maker or governing body of one game and yet be the player of another. Business-owners are players in a game whose rules are made by the national government. In turn, each business-owner sets up rules for promotions in his firm, thereby creating a game played by his employees. Yet other games may be played by this same person, including "Strike," the union-management game (Section 5.2.1) and variants of "Not Asking for a Raise" (Section 2.4).

1.2 Games This Book Is Not About

The word "game" means different things to different people. In particular, it may make people think of parlor games, simulation games, and "games people play." These other concepts are emphatically *not* the subject matter of this book, but in this section we will take a look at their relationship to our definition.

Parlor games, although they fit our definition of games as interactive-decision structures, are not really our main concern. We do use them as introductory examples, for simplicity, but they are of limited relevance to real situations because they are zero-sum (see Section 1.1). Of course two real people may play chess and both enjoy it, so that in their social decision to play they both "win" some enjoyment. But within the framework of chess itself, the players are irreconcilable adversaries.

Games People Play (1964) was an enormously popular book in which the author, Eric Berne, characterized some unsatisfying social behavior sequences and called them games. The particular social situations that Berne describes could certainly be profitably examined within our framework. However, we will see that his approach and objectives differ greatly from ours.

Considerable thinking has been done in "think tanks" about such global concerns as war and diplomacy, using simulation games as a means of analysis. Games simulating the many complexities of business and society have been used in both research and teaching. However, the simulation approach is significantly different from that of game theory.

How do these other approaches differ from game-theoretic analysis? Recall that we are concerned with interactive-decision structures: only those who make *decisions* are considered bona fide players, and players *interact* if their decisions affect each other's outcomes. We are interested in decision *structures* in the sense that a plan of action may consist of several decisions, some of them contingent on what other players may decide to do.

Which of the following are games in this sense? That is, which of them are interactive-decision structures?

- ☐ Chess
- ☐ Poker (say, five-card draw)
- ☐ Football
- ☐ Marriage
- ☐ "If It Weren't for You" (one of the "games people play")
- ☐ The pollution crisis (or situation)
- ☐ SIMSOC (a simulation game used in sociology classes)

The answer is all of them, but we shall not be equally interested in them all.

The first three, chess, poker, and football, are games in the more common sense of artificially contrived contests for physical or intellectual pleasure. Because we wish to focus on real (uncontrived) social phenomena, we shall not be directly concerned with such games. However, because they are familiar and have precise rules, contrived games can be useful examples of such notions as probability and information. For example, poker begins with a random event called shuffling, which makes probability calculations a crucial part of that game. In chess, on the other hand, once it is determined who goes first, probability plays no role. With respect to information, in the game of "I'll choose you for it" no player is allowed to know the others' choices, while in chess everything is in the open. (In "I'll choose you for it," we each put out one or two fingers and you win if the sum is even but I win if it's odd.)

Although marriage is an interactive-decision structure, it is too complex a subject to be reduced to a single, simple, readily grasped set of rules—as anyone who has tried it knows. Certain *aspects* of marriage can, however, be expressed in simple terms. One type of situation that occurs frequently *within* a marriage is described in the game-theory literature under the rubric "Battle of the Sexes." It involves two individuals who want to go out together for the evening, but, unfortunately, have different ideas about where to go. To keep things simple, it is assumed that there are only three possible outcomes: the preferred place of each person or a stalemated, unhappy evening at home. Each person is taken to have two possible stances, yielding or unyielding. Also in the interest of simplicity, nothing is said about the personalities of the individuals or the past or future of their relationship. So there is a twofold simplification: first, a specific situation (an evening's entertainment) rather than a whole marriage is analyzed, and second, only the broad outline of the situation, the decision structure, is considered. (A better name for this game is "Let's all have fun doing it my way," since the description can apply to all sex combinations and any number of players.)

This process of singling out the decision structure of a situation is a principal focus of this book, and will largely determine which real-world situations we choose to look at. For every situation we want to end up with a simple structure because complex ones are hard to analyze. Ultimately we hope to be able to say something about how real individuals (or nations, etc.) are likely to act within the decision structure and how their deliberations reflect the potential choices and results. To end up with an analysis that gives us this kind of insight into real situations, we must choose coherent topics and make sensible simplifications. This is no easy goal, and the reader is invited to improve upon the examples in this book.

Next on the above list we find "If It Weren't for You" from Berne's *Games People Play*.

> Mrs. White complained that her husband severely restricted her social activities, so that she had never learned to dance.... [She later discovered] that she had a morbid fear of dance floors... [By picking a domineering husband, she had put herself] in a position to complain that she could do all sorts of things "if it weren't for you".... [When she and her friends] met for their morning coffee, they spent a good deal of time playing "If It Weren't for Him." [p. 50]

With this and other games, Berne does a masterful job of selecting key sequences of "moves" in social interaction and analyzing the "payoff" or psychosocial advantage that motivates people to behave the way they do in specific situations. Moves and payoffs will be important notions to us, too. Berne, however, usually analyzes an interactive situation from the standpoint of a principal player who is set in his ways, and focuses on a rigid sequence of events rather than the range of interactive possibilities. Thus, Berne's players do not seem to make decisions because they are blind to their options. Finally, dealing as he is with what he calls social psychiatry, Berne invokes certain restrictions. He says that "every game ... is basically dishonest, and the outcome has a dramatic ... quality" (p. 48). Moreover, "some implicit or explicit judgment is passed on the 'healthiness' of the game studied" (p. 51). This psychiatric viewpoint is not relevant to our interdisciplinary approach. Nevertheless, the reader may find it intriguing to reanalyze Berne's games within the framework presented here.

Like marriage, pollution is not a single coherent decision structure, so it is important to select a manageable subtopic for analysis. Decisions about whether and how much to pollute are made at many levels of social organization, including individuals, businesses, states, and nations. For example, factories discharging chemical waste into a river may be challenged by individuals downstream who want to use the river for recreation. In this game (interactive-decision structure) the various participants are clearly on an unequal footing, since factories can make water unfit for swimmers but not vice versa. On the other hand, when cities discharge sewage into a lake, each city's decision has a harmful effect on every other (at least in the long run). Thus there can be different kinds of structures and different kinds of players, depending on which aspect of pollution one chooses to look at.

In SIMSOC, the last item on our list, the society simulated by the players

> ... has certain basic groups—economic, political, mass media, and others—but it has no government. The game starts with a collection of diverse interests, and it is up to the players to create institutions (especially governmental ones) and a normative order as the need arises. To do this, they meet with other individuals of like interest, bargain with individuals who have different interests, travel about to influence players in other regions, threaten others with arrest and confiscation of their resources, and generally "wheel and deal." [Gamson, 1972, p. 55]

SIMSOC requires 20 to 60 players and thus is smaller and simpler than real societies (which take millions of players). Nevertheless, SIMSOC is very complex, involving all sorts of social and economic indicators and political structures. It is this great complexity that distinguishes SIMSOC and other simulation games from the relatively simple interactive-decision structures we shall deal with. If a simulation game provides realistic decision-making roles and specifies significant options and outcomes, it will give its players a feel for how a society functions. It will not, however, provide analysis or prediction.*

1.3 The Interdisciplinary Nature of Game Theory

The situations used in this book as examples of interactive-decision structures are the subject matter of a variety of disciplines: social psychology, economics, political science, and international relations. The unifying element in our study of such diverse situations is the individual decision-maker. When only a few actors are involved, such as in a couple or family or in relations between two or three superpowers, the effect of a single decision by one actor may be crucial in determining the outcome and will almost certainly affect all parties. For example, the threat of divorce or nuclear warfare obviously has important consequences for the threatening party as well as the threatened one. On the other hand, in a large group, such as the residents of the Los Angeles air basin or the supporters of a presidential candidate in a primary, a single actor's decisions

* The reader interested in simulation games will find Inbar and Stoll (1972) and Shubik (1975) of interest.

have little effect on any other individual. Thus, the decision of one commuter to take the bus instead of driving her car or a single $25 contribution will probably go unnoticed. These considerations suggest that the *number of participants* is an important property of situations.

Our analyses will proceed in terms of structural properties, such as the number of participants, rather than according to subject matter. Thus, rather than leaving families to sociologists, nuclear deterrence and the Strategic Arms Limitation Talks to international relations specialists, transportation to city planners, geographers, or economists, and election campaigns to political scientists, we shall consider all forms of social interaction, classifying them in terms of decision structures and analyzing the behavior of the participants in terms of those structures.

If situations from different traditional disciplines are lumped together, so much the better. If the cold war resembles a lover's quarrel, let us ask in what way. The typical lover does not have a secretary of defense or an opposition party, but that may not keep lovers from escalation and de-escalation of hostilities. On the other hand, situations that seem to be related and that have traditionally been lumped into a single discipline may have to be unlumped. An example is legislative and electoral politics, both of which are traditionally the province of political science. Although it is important to study these two phenomena together, since they ultimately affect each other, it may be that their interactive-decision structures are vastly different. If the discipline of political parties is so strict that all legislators in a party vote together, then the parties themselves can be viewed as decision-makers and legislative politics can be analyzed as a game with few players (since there are usually only a handful of political parties active). In contrast, voters are numerous, so that voting is a game with many players. Moreover, parties (through their leadership) can communicate with each other more readily than large numbers of voters; they are also required to make interactive decisions more frequently. These two interactive-decision structures may differ in other aspects as well, but perhaps the point is clear enough: the character of interactive-decision-making processes transcends traditional academic boundaries.

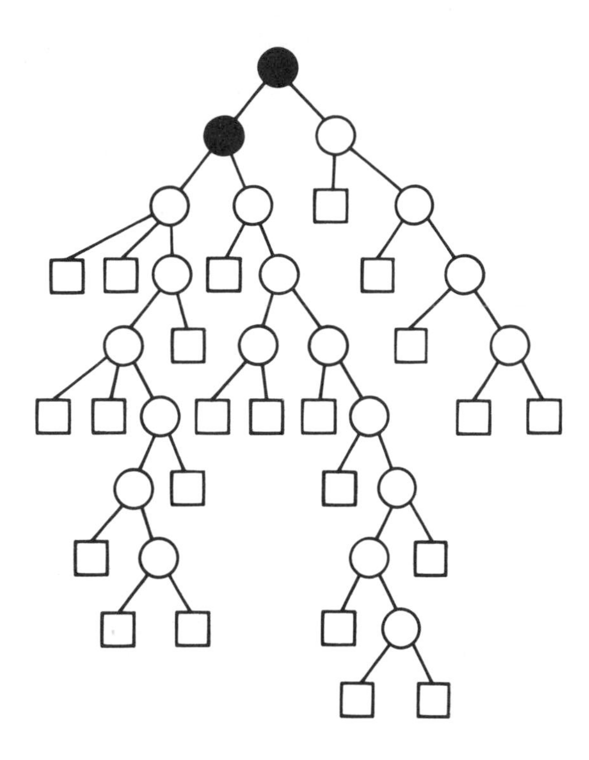

2 The Language of Games

A game-theoretic analysis has four basic ingredients: the players, their options, the possible results, and the players' preferences among those results. These four items are neatly represented by either of two kinds of diagrams, called trees and matrices.

Interactive-decision structures can be described using words or trees or matrices. In addition to explaining what the latter two are, this chapter will show how to translate among all three. Sections 2.1 and 2.2 use simple, artificially contrived games to show how verbal descriptions are converted into matrices and trees. Two rather complex real-life interactive-decision structures are similarly translated in Section 2.3, one into a matrix, the other into a tree. Translating real-life examples requires simplifications that must be made with care if the essence of the situation is to be retained. The final section relates the two theoretical structures, trees and matrices, by translating one into the other. (Additional ways of representing games appear in Chapters 5, 7, and 8.)

2.1 Verbal and Matrix Descriptions

A game can be described in words. Consider the following example:

- **Game 2.1 Matching Pennies**
In this two-person game each player takes a penny and places it either head-up or tail-up and covers it so the other player cannot see it. Both players' pennies are then uncovered simultaneously. One player is called Matchmaker. This player gets both pennies if they show the same face (both heads or both tails). The other player is called Variety-seeker. This player gets the pennies if they show opposite faces (one head and one tail).

The entire description of Game 2.1 can be replaced by a simple chart or matrix containing all the key information. Thus in Matrix 2.1, Matchmaker may choose the top row (heads) or the

		Variety-seeker	
		Heads	**Tails**
Matchmaker	**Heads**	+1	−1
	Tails	−1	+1

Matrix 2.1 "Matching Pennies."

bottom row (tails). Variety-seeker may pick the left or right column. Their two choices together determine a box or cell of the matrix. For example, if both pick tails, the result is the lower-right cell, in which we find the number +1. This number is the *payoff* to Matchmaker, indicating that Matchmaker wins 1¢ in this case (because he keeps his own penny, his net change in wealth is +1).

This parlor game is a zero-sum game, since money is neither created nor destroyed. The matrix has been made from Matchmaker's point of view, so that, for example, it has −1 when he loses. It is conventional to give payoffs only for the row-chooser in a zero-sum game, and this has been done in Matrix 2.1. Since Matchmaker's loss is Variety-seeker's gain, one could deduce

Variety-seeker's payoffs simply by replacing plus signs with minus signs and vice versa in Matrix 2.1.

Notice that the matrix clearly displays the names of the players, the options available to them, and the way in which those options can interact to give results. In addition to players, options, and results, the fourth specification, preferences, is needed. For the time being it will simply be assumed that winnings in money are an exact measure of how well satisfied each person is. This topic will be discussed in detail in the next chapter.

Game 2.2 Matching Nickels

This game is played the same way as "Matching Pennies" but the stakes are five times as high.

It will probably be useful to the reader to draw the matrix of Game 2.2. This time put the options of Variety-seeker in the rows and fill in payoffs from her point of view, that is, plus when she wins and minus when she loses. (It is conventional to fill in the matrix from the viewpoint of the actor represented in the rows.)

Game 2.3 Old Shell Game

In this game Showman puts a pea under one of three identical shells, called A, B, and C. Carnival-goer must then guess which shell has the pea. The payoffs to Carnival-goer, shown in Matrix 2.2, assume that the fee for the game is 25¢ and that the prize is worth 65¢, so that winning represents a net gain of 40¢.

Payoffs to Showman are the opposite of those shown in Matrix 2.2; that is, he gets +25¢ when Carnival-goer gets −25¢, and −40¢ when Carnival-goer gets +40¢. We ignore the possibility that Showman may have bought the Kewpie dolls he gives as prizes at wholesale prices, or that a Kewpie doll is not worth its 65¢ retail price to Carnival-goer since she would never have bought one. We also ignore her satisfaction in winning and the benefits of demonstrating her skill and perspicacity to her boyfriend. Thus, as in "Matching Pennies," one player's loss is always the other player's gain and so,

		Showman		
		A	B	C
Carnival-goer	A	40	−25	−25
	B	−25	40	−25
	C	−25	−25	40

Matrix 2.2 "Old Shell Game."

no matter what happens, the sum of the payoffs is zero. When Carnival-goer wins, the sum of the payoffs is $(+40) + (-40) = 0$; and when she loses, it is $(-25) + (+25) = 0$. Therefore, this is a zero-sum game.

If the sum of payoffs to the players is always equal to some particular number (not necessarily zero), the game is "constant-sum." Such games present little additional strategic interest, as you will see by working through exercise 3. If, as is typical of games that model real life, the sum of payoffs depends on the outcome, then we have a "variable-sum game."

Zero-sum games are particularly easy to work with, and many of the ideas that are useful for all games can be most easily illustrated with them. Moreover, even where they differ from other games, they form a useful basis of comparison. However, it must be stressed that most games derived from real-life situations are not zero-sum and the use of "zero-sum thinking" in situations that are not zero-sum can be counter-productive. Many interesting mathematical results have been obtained that apply only in the zero-sum case. While it seems a shame not to apply these results more generally, it is also quite pointless to try to use them where they are meaningless. We shall return to this topic somewhat later, after having looked at some games that model real life.

Notice in Game 2.3 that although Carnival-goer chooses *after* Showman, she gets no advantage from doing so since she is not allowed to see what Showman picked. Therefore, the two players could just as well make their choices simultaneously, as in "Matching Pennies." Another way to look at it is to say that although Carnival-goer goes second, she has *imperfect information* (in fact, no information at all) about Showman's prior choice.

The following is a well-known zero-sum game; exercise 4 asks you to translate it to matrix form.

● **Game 2.4 Scissors, Paper, Rock**
Each of the two players simultaneously puts out two fingers (representing scissors), a flat hand (paper), or a fist (rock). Scissors cuts paper; paper covers rock; rock breaks scissors. Whoever can "cut," "cover," or "break" the other player wins one point from the other player, who loses one. If both players put out the same thing, each gets zero.

2.2 Tree Descriptions

In the preceding section we saw that a verbal description could be summarized in the form of a matrix. In doing so, we were careful to use games in which players in effect make choices simultaneously and each player makes only one choice in the entire game. The following game has neither of these conditions.

Game 2.5 Pile of Four Stones

Two players, A and B, take turns removing stones from a pile. The pile has four stones at the start and each player in turn may remove one or two stones. Whoever gets the last stone wins; A goes first.

This game may be put into the form of a tree that shows the choices available to each player at each turn (Tree 2.1). The game

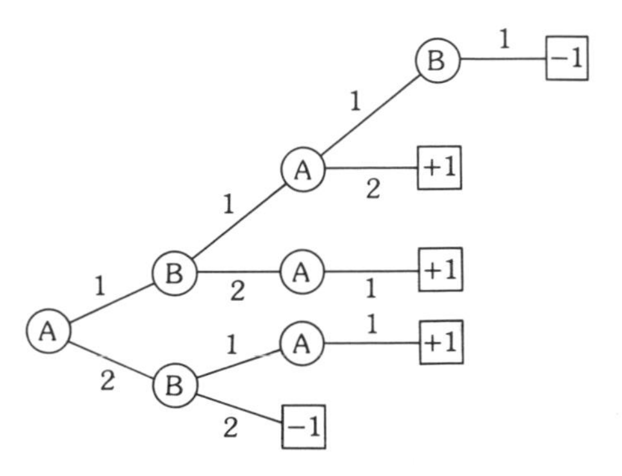

Tree 2.1 "Pile of Four Stones."

begins at the left side and moves to the right. The numbers 1 and 2 on the branches refer to the choice of removing one or two stones. Also included on the tree are the payoffs to player A; a win is shown as $+1$ and a loss as -1. Of course, a loss to A is a win for B, so B's victories are represented in this matrix by -1 and his losses by $+1$.

Game 2.6 Boxes

Player A goes first and places a marker in any of five boxes (U–Y). Then player B places a marker in any box that shares a side with the box chosen by A. Players alternate in this way, always choosing a new box adjoining the one just chosen by the other. If there is no such box, no further moves are made. The last person to place a marker is the winner.

U	V	
W	X	Y

For example, if A picks X, then B may pick W, V, or Y, but not U. Suppose B picks V; player A must then pick U, and after that B must pick W, thereby winning the game since all boxes are now used except Y, which does not share a side with W. Try to draw the game tree for Game 2.6. Part of that tree is supplied in Tree 2.2.

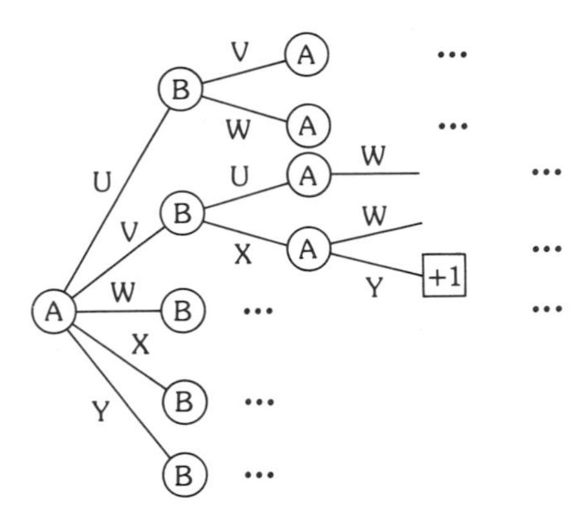

Tree 2.2 Part of the tree for "Boxes."

Why use trees when we already have matrices available as a means of diagramming games? Notice in both Games 2.5 and 2.6 that (1) players have more than one move; (2) moves are not made simultaneously; and (3) the results of all previous moves are always made public so that these are games of "perfect information." The trees enable us to see the sequence of possible choices available to each player in turn. There is no obvious way to show this sequence in a matrix.

Game 2.7 Glass-Shell Game

A version of the "Old Shell Game," but with transparent shells. Showman "hides" the pea, and then Carnival-goer "guesses," but of course it is no guess since she can see what the right choice is.

It will be useful for the reader to draw a *tree* for Game 2.7. This game is identical to the "Old Shell Game" (Game 2.3) except for the information condition, i.e., Game 2.7 has perfect information while Game 2.3 had imperfect information. Thus, by letting one of the players go second and informing her of the other's move, we have formed a tree game from a matrix game. In the final section of this chapter we shall go in the opposite direction, forming matrix games from tree games, without even changing the information conditions. That is, a tree game will be translated, not modified, into a matrix game. The matrix game will therefore be exactly the same game as the original tree game.

The "Glass-Shell Game" is too simple to hold much interest. Carnival-goer has an unfair advantage and it is easy to see how she can exploit it. Less obvious is the fact that player A in "Pile of Four Stones" also has an unfair advantage. To analyze this game in detail, consider its tree (Tree 2.1). Perhaps you noticed that when there is only one stone left on the pile, there is really no decision-making left to be done. The player whose turn it is must simply pick up the remaining stone and thereby win. We may imagine that a referee can do this for him automatically and that the game has, for all intents and purposes, really ended one step earlier. Fixing up the tree to reflect all this, we obtain Tree 2.3.

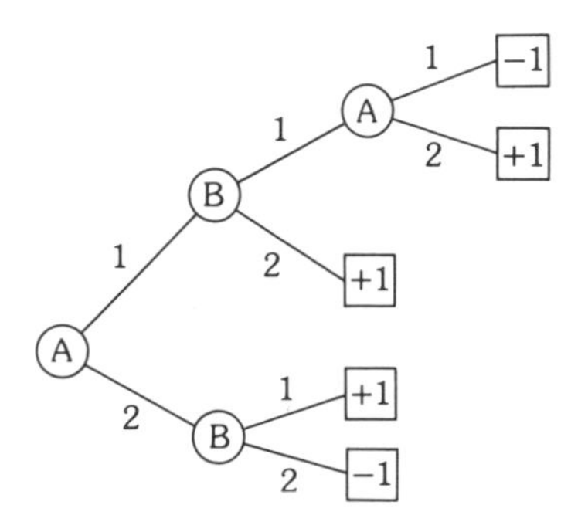

Tree 2.3 Refinement of Tree 2.1, "Pile of Four Stones."

In Tree 2.3 it is now particularly clear, for example, that if A takes one stone and B then takes two, A wins (the referee will pick up the last stone for A). It is also pretty clear that certain choices would be rather foolish. For example, find the point in Tree 2.3 that would be reached after A picks one and B picks one. It is now A's turn and he will win or lose on this choice. In effect, the tree says to him at this point, "You may pick two and win or pick one and lose; which do you pick?" We assume that players are trying to win, so there is no need to ask such a question and we can simply assign the win to A immediately after B's move. Similarly, if at the start of the game A were to pick two, then B would have a sure win available and we would merely assign the win to B on the basis of A's move. Fixing up the tree to reflect all this, and remembering that a win by B is represented by −1 (payoff to A), we get to Tree 2.4.

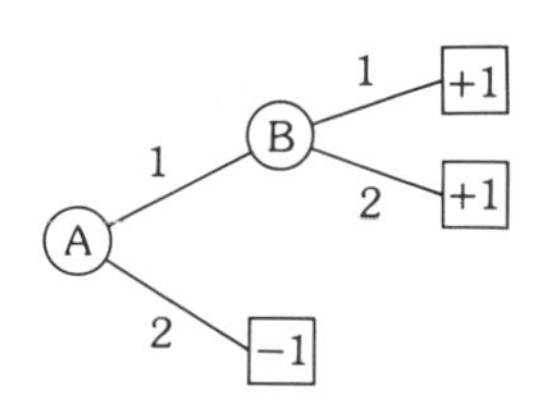

Tree 2.4 Refinement of Tree 2.3.

In Tree 2.4 player B is confronted by a peculiar choice. There seem to be two options available, but both lead to the same result, a payoff of +1 (a win for A). Since no real choice is involved, we can simply let the referee declare the outcome without consulting B. Fixing up the tree to reflect this aspect of the situation yields Tree 2.5.

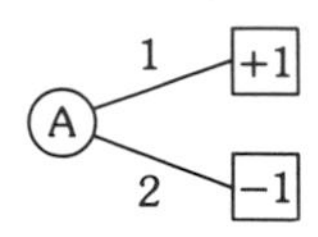

Tree 2.5 Refinement of Tree 2.4.

Now we see that the game, as represented in Tree 2.5, simply amounts to asking A whether he would like to pick one and win or pick two and lose. In other words, *by picking one on his first move, A can insure victory.* What A does on his second move in order to complete his victory will depend on what B has done in the meantime, but it will be an easy choice in any event. You should now write down exactly what the possibilities are and what A should do in each case.

In summary, the tree has been repeatedly simplified by eliminating the following:

☐ **1** Places where a player only has a single option and thus does not really have any decision to make (in Tree 2.1).

☐ **2** Easy decisions, where a player must choose "immediate win" in preference to "immediate loss" (in Trees 2.3 and 2.5). These are true decisions and illustrate the idea of rational choice in its simplest form.

☐ **3** Pseudo-decisions, where a player has more than one option but they all lead immediately to the same kind of result, i.e., win or loss (in Tree 2.4).

The method we have just used, working backward through the tree to show that one of the players can insure a victory, can be used for any game tree, though the results will not always be the same. In our example it turned out that player A (who went first) had a sure win, but in other games it may be that player B (who goes second) will be able to insure a win. Who can insure a win in the following games? In each case, how?

● **Game 2.8 Pile of Five Stones**

● **Game 2.9 Pile of Six Stones**

Game 2.10 Pile of Nine Stones
Players may remove one, two, or three stones at each turn.

(Answer for Game 2.9: player B can win if she always does the opposite of what player A has just done. You can show this with a tree, or note that three stones will be taken in the first two moves and three more in the next two.) Although it is A who can insure a win in some games and B in others, one general conclusion always holds, as shown by the fact that we can always work backward all the way to the start of the tree: *In a two-person game of perfect information the outcome is strictly determined.* In games such as 2.5, 2.6, 2.8, 2.9, and 2.10, where payoffs are simply "win" and "lose," strict determinacy means we can be sure that one of the players will be able to guarantee himself a win and the other can do nothing to stop it. However, there may be several different payoffs available in a game, as the following example shows.

Game 2.11 Variant of Four Stones
Same as Game 2.5 except that the winner gets 2¢ from the loser if two stones were taken on the first move of the game and 1¢ if one stone was taken.

You should now draw the tree for Game 2.11, including payoffs from A's point of view. Possible payoffs for A in this game are +2, +1, −1, and −2. The strictly determined outcome here is +1, since A can guarantee a payoff of at least +1 (how?) and also B can guarantee a result no higher than +1 (a loss of 1¢ by B).

2.3 The Variable-Sum Games of Life

The games in the preceding two sections are all unambiguous. The translation of each from a verbal description to a description in formal terms leaves little room for interpretation or opinion. For "Matching Pennies," for example, once we agree to use a matrix and make a few arbitrary decisions, like whether to put the label "heads" on the top or bottom row, there can be little argument over which payoffs are +1 and which are −1, or whether one of the payoffs is really +3, and so forth. As long as cents (or dollars or "points") are the units of the game in both verbal and formal description, the translation from one description to the other is automatic, provided the verbal description is thorough. The games in the last two sections were also all zero-sum.

In this section we will try to characterize real-life situations in the form of games. Assigning payoff values will not be a straightforward procedure. Outcomes will typically not be simply the winning of money or points but may instead concern such weighty matters as love, war, and peace. Even though it is often difficult to assign exact numerical values to real-life outcomes, it is easy to see that certain games are not zero-sum. For example, certain moves by nations possessing hydrogen bombs would result in nuclear holocaust, and such a result is clearly worse for both (all) nations than the prior situation. But in a zero-sum game a change for the worse in one player's circumstances must be accompanied by an improvement in the other's.

To make things more precise let us formulate the foregoing situation as a game. In doing so, we will make some drastic simplifications and disputable assumptions. First we introduce the game itself, then turn to its relationship, if any, to real life.

● **Game 2.12 Nuclear Chicken**

Two players, A and B, each must choose alternative C or alternative D. The resulting payoffs are given in Matrix 2.3. Players may not communicate.

		Player B	
		C	D
Player A	C	0, 0	−2, 1
	D	1, −2	−10, −10

Matrix 2.3 "Nuclear Chicken."

As can easily be seen in Matrix 2.3, there are two payoffs in each cell of the matrix. Those of the row-chooser, in this case A, are placed before the comma in each cell and those of the column-chooser come after it. For example, if player A picks D and player B picks C, then the result, shown in the lower-left cell, is that A gets +1 and B gets −2.

I have given this game a very provocative name without providing any justification for it. But before discussing the game in terms of nuclear confrontation, we can see just by looking at the numbers in Matrix 2.3 that a serious dilemma inheres in them. For the moment, imagine that two mature, intelligent beings, yourself and somebody else, have been placed in separate rooms and each asked to choose C or D, without consulting the other player. Assume that the units of payoff are 100-dollar bills so that, for example, −2 stands for a loss of $200.

You are player A. What do you choose? What do you think B will choose? Is there any reason why B would choose differently than A? If both players are determined to avoid a loss of $1,000 (ten 100-dollar bills), then both will choose C. But if you think the other player will choose C, then by picking D you can get $100 (+1 in the matrix). So there is a temptation for either player to pick D, hoping the other will not. But if both pick D, they both suffer their worst possible outcome.

Now replace A and B by the Soviet Union and the United States and let the alternatives C and D stand respectively for accommodation and aggression with nuclear weapons. Assume that if both sides choose accommodation, then the status quo will continue; this yields neither gain nor loss, hence a payoff of zero seems appropriate. If one side is aggressive (D) and the other accommodates, then the accommodating side would suffer strategic and perhaps trade and diplomatic losses (−2), while corresponding gains would accrue to the aggressor, though in lesser amount (+1) because the resulting increase in world tension might generally interfere with trade and diplomacy. If both sides choose confrontation (with no willingness to accommodate), then the result is taken to be a nuclear exchange, presumably by far the worst outcome for each (−10).

Many questions are left unanswered in this presentation. What happens after the "nuclear exchange"? Is there any way to stop short of all-out war once the nuclear threshold is crossed? What of alternatives other than the two mentioned? In the Cuban missile crisis, the world's closest brush with nuclear hostilities so far, the U.S. could have tried to invade Cuba by sea, dropped bombs, done nothing, or offered to withdraw missiles from Turkey in exchange for withdrawal of Soviet missiles from Cuba. There were many other alternatives and the Soviet Union also had more than two available courses of action. What of communication? In Game 2.12 players may not talk, but during the Cuba confrontation there were messages in both directions, one of them in fact containing a Soviet offer of withdrawal from Cuba in exchange for U.S. withdrawal from Turkey. What of other countries? Should China be included? Europe? Each nation that has a bomb? Only those that have a sophisticated delivery system? We shall return to this game and some of these questions later (in Sections 4.2.3, 4.3.2, and 7.3.2).

Rolling back the clock a few centuries, let us consider the contest between King Charles I of England and the Independents. In the winter of 1648–1649 the turbulent reign of Charles I was coming to a troubled end. The king had consorted with foreign powers to make war on his own people in a last-ditch effort to save his rule. In doing so, he had grossly violated a treaty he himself had signed, thereby convincing many that the nation could not be at peace while he lived.

We shall formulate a game from the events in the closing months of Charles's reign. The analysis will be of interest in its own right in clarifying some puzzling aspects of the events. Equally important, it will provide several contrasts with the foregoing example of "Nuclear Chicken." It takes place in the seventeenth rather than the twentieth century; the players are political factions rather than nations; and the game will be expressed as a tree rather than as a matrix, to emphasize the sequence of the decision-making. The considerable historical detail presented below will show that the game we formulate is a much abbreviated version of the real situation. The game is, after all, a *model,* and you will have to judge whether in this model the gain in clarity is worth the loss of historical detail.

We find four parties of political interest: the king; the Independents, led by Oliver Cromwell, who controlled the army but were a minority in Parliament; the Presbyterians, who held a parliamentary majority; and the Levellers, whose ideas of opportunity, justice, and widening of the franchise had considerable support from the rank and file of the army. None of these groups was of a single mind in its decisions, and a more detailed description would distinguish various subgroups and individuals. However, we shall move in the opposite direction, focusing attention on the decision possibilities of only two of the parties, the Independents and the king. The following synopsis relies on *A Coffin for King Charles* (Wedgwood, 1966); page references are to that book, in which may be found sources for the quotations from historical figures.

In November 1648 the Independents, in control of the victorious army, were in a position to impose military rule on the nation. However, for a variety of religious, philosophical, and practical political reasons, they sought to find at least some semblance of a legal basis on which to establish their control. They therefore proposed to the king to restore him "to a condition of safety, honour, and freedom" (p. 20) in exchange for a regularly elected Parliament that would, in effect, run the country. Upon the king's refusal, the army sent a Remonstrance to Parliament asking that the king be brought to trial, thereby setting in motion an unprecedented series of events.

> ... Kings had been killed before, had fallen victims to conspiracy, had been deposed, had been murdered. The grandmother of Charles I, Mary, Queen of Scots, had been tried and executed; but not while she was a reigning Queen,

not in her own country or by her own subjects.... King Charles was brought to trial by his own people, under his title as King—an act which defied tradition and seemed to many a fearful blasphemy against a divinely appointed sovereign. [p. 1]

Thus began the series of interactive decisions that we represent as a game in Tree 2.6. Having determined to bring the king to

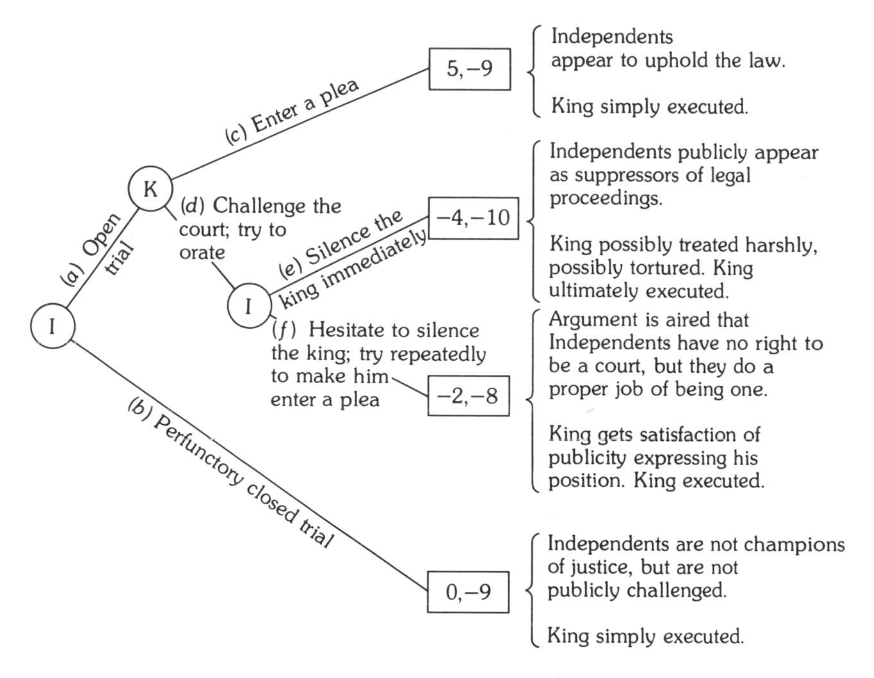

Tree 2.6 Game tree representing the conflict between Charles I of England (K) and the Independents (I). Payoffs before the comma are for the Independents; those following the comma are for the king.

trial, the Independents had to decide how to go about this unprecedented procedure. In our game tree this decision is shown with two alternatives, an "open trial" or a "closed trial." Of course there are gradations of openness, but by and large the trial was conducted in quite a public manner, all things considered; branch *a* was thus chosen.

The king's captors had hoped that by conducting an open trial they would make clear to God and their countrymen that they were men of conscience, fairness, and law. They must also have realized that in doing so they were providing the king with a public forum.

Presumably, they anticipated that he would plead innocent and give a few excuses for his actions. (This possibility is shown as branch *c* in Tree 2.6.) The Independents in turn would have the opportunity to recite their litany of grievances against him, thereby proving the justness of their actions.

But the king would plead neither guilty nor innocent. Instead he attacked the legal basis of the court, which was made up mostly of members of the House of Commons, the lower house of Parliament: "I would know by what power I am called hither." After being rebuked a couple of times for his intransigence, the king nevertheless returned to his theme:

> I do not come here as submitting to the Court: I will stand as much for the privilege of the House of Commons, rightly understood, as any man here whatsoever. I see no House of Lords here that may constitute a Parliament.... [p. 122]

This was a telling point, since the House of Lords had refused to participate and the House of Commons, which was in any case no kind of court, had been forcibly "purged" of many of its members who would not countenance the trial. The king's decision is represented by branch *d* in Tree 2.6.

Such uncooperative behavior from a man who was, after all, physically at the mercy of the court can perhaps be explained in terms of his convictions, but the strategic structure of the situation is also relevant. For after this perhaps surprising detour in the decision tree, it is the Independents' move. Shall they terminate the proceedings (branch *e*)? And if not, how shall they silence the king or force him to keep to their intended procedure?

> ... it was not possible either to examine witnesses or to make out a public case for the prosecution if the accused stood mute or pleaded guilty, for in that case—logically enough—no such demonstrations were required by English law. Therefore the conduct of the King destroyed a principal purpose of the trial. Certainly he could be taken as guilty and sentenced to death; but he could not be *proved* guilty for all the world to see. [p. 125]

The dilemma of the Independents (represented by the choice between branches *e* and *f* in Tree 2.6) is that although they can make the king pay for his intransigence, it is not fully in their own interest to do so. The hastier they are in silencing him, the more they appear to be simply a kangaroo court. Thus, it is only after much delay and internal disagreement that they finally hustle him out of court and away to his execution.*

At the outset we pointed out that we were leaving the Presbyterians (with their parliamentary majority) out of the decision-making. The resulting model has allowed us to focus on the interactive sequence of decisions by the king and the Independents. But by leaving out the Presbyterians we have ignored a rather important part of the story, one with an entirely different kind of decision structure. By including the Presbyterians we increase the number of players from two to three, and thereby introduce the possibility of coalition formation. At one point in the conflict it seemed that the king and the Presbyterian majority in Parliament might agree to share power between them, leaving the Independents in the army with no political or legal action. The threat to the army of being left out of a successful coalition was expressed by Independent Thomas Harrison.

> We fully understand that the Treaty betwixt the King and Parliament is almost concluded upon; at the conclusion of which, we shall be commanded by King and Parliament to disband, the which if we do, we are unavoidably destroyed... and if we do not disband, they will by Act of Parliament proclaim us traitors, and declare us to be the only hinderer of settling peace in the nation. [p. 19]

In this particular three-party encounter (a three-player game) no coalition actually formed. The study of coalitions from the viewpoint of game theory is concerned with which pairs or groups of players can profitably unite, how they can or should or will divide the fruits of their bargain, and whether they will keep their word or whether outsiders will be able to tempt them into new coalitions by offering better deals. This topic is treated in Section 8.1.

* In this account we have focused on political-strategic considerations, leaving out socio-economic and religious factors. A proper account should stress that England was to be the vanguard of European commerce for the next two centuries, and that the rising middle class found the capricious rule of its monarch highly inconvenient to the free flow of goods and money. One must also take note of the continual claims on both sides that they were only doing their duty as Christians. Charles, purportedly king by divine right, was nevertheless referred to by Cromwell as "this man against whom the Lord hath witnessed." A possible interaction between economics and religion has been argued by Max Weber, who explores the convenient compatibility between the Puritan ethic and the growth of commerce.

Of the two real-life situations described in this section, one (the Cuban missile crisis) was represented as a game matrix and the other (Charles I versus the Independents) as a game tree. The choice between tree and matrix is influenced by the intrinsic nature of the situation being analyzed, but it also depends on our own analytic interpretation of what is important or of special interest in the situation. Also left to the analyst's discretion is the choice of which parties to exclude from consideration, in the name of simplification. Finally, note the difficulty in stating what the payoffs should be. We have seen, however, that payoffs are not entirely arbitrary, and that "Nuclear Chicken" in particular is clearly not a zero-sum game.

2.4 Translating Trees into Matrices

So far trees and matrices have been used for different kinds of situations. Matrices have been used to represent simultaneous choices for two players, while trees have allowed us to express a succession of moves, usually with a player knowing everything that has been done up to the time of a particular decision. Since a principal objective of game analysis is to compare different situations, it would be most helpful if we could express all games in a single format. This can in fact be done, by translating trees into matrices. This procedure will be easier to follow if we digress briefly to introduce a particularly simple tree.

The simplest possible game tree is one in which (1) the first player chooses from among only two possible alternatives; (2) one of these alternatives terminates the game while the other alternative presents a decision for the second player; and (3) this latter move presents only two alternatives and either of them ends the game. The tree for such a game will look like Tree 2.7. In this tree X, Y,

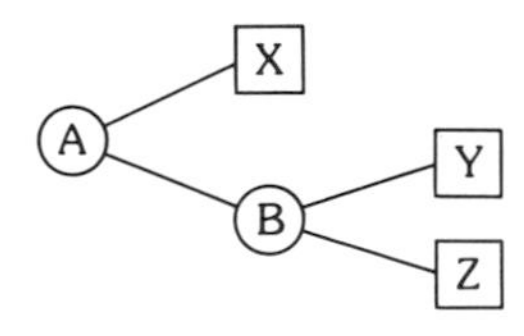

Tree 2.7 The simplest possible game tree.

and Z stand for outcomes or states of affairs in the world. Each state of affairs will yield a payoff to each player. Thus X might be (6, −3), meaning a payoff of 6 to A and −3 to B.

Little games of the sort represented in Tree 2.7 crop up all the time in real life. In fact, sometimes they are going on right before your eyes but without your knowing it. Such games are, in a manner of speaking, invisible. For example, consider the game of "Not Asking for a Raise," shown as Tree 2.8.

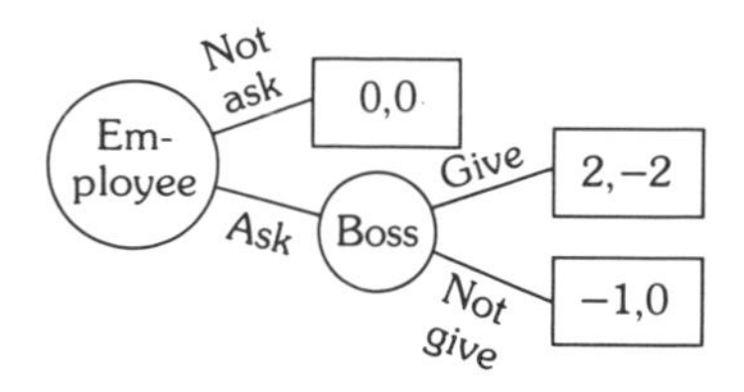

Tree 2.8 "Not Asking for a Raise."

As you can see, Tree 2.8 is an example of the general form in Tree 2.7. If the employee never asks for a raise, the status quo continues. This outcome is associated with the payoffs (0, 0). A denied request, we may imagine, does not affect the boss but puts the employee in a sensitive position, somewhat less pleasant than the status quo, hence the payoff −1. (In the interest of simplicity we have ignored the fact that the employee may then quit, be fired, or sow dissension among the other workers, who may in turn react one way or another to whatever happens.) If the raise is given, the employee is financially better off at the boss's expense, hence the payoff (2, −2).

Why is this game invisible? To answer this question we ask what the boss will do if faced with a decision. Given the payoffs that we assumed in Tree 2.8, she will surely refuse the request. Knowing this, the employee never asks. Thus nothing visible ever happens, unless you are the employee's confidante and are told, "Gee, I sure would like a raise but of course I'd never get it so why bother asking."

It is now time to translate this tree into a matrix. To do so, imagine that the boss is going away on vacation and must leave behind instructions in a sealed envelope for her assistant, telling what to do in any possible contingency that may arise. In this particular simple game the only contingency that will require a decision is if the employee asks for a raise. Thus the boss must put in the envelope one of the following instructions: "If employee asks for a raise, say no." Or, "If employee asks for a raise, say yes." Once this little ploy has been carried out, it is possible to imagine both players choosing simultaneously. The game would thus be properly represented by Matrix 2.4. Imagine that the referee asks the employee for a decision and simultaneously takes the sealed envelope: If the employee's choice is *not* to ask, then the envelope is returned unopened and the payoffs are (0, 0). Notice that this pair of payoffs

		Boss No	Boss Yes
Employee	Ask	−1, 0	2, −2
	Not ask	0, 0	0, 0

Matrix 2.4 Matrix translation of Tree 2.8.

appears in both cells of the bottom row. If the employee *does* ask, then the envelope is opened and the instruction in it is carried out.

More complicated trees can be translated into matrix form. The key to the conversion is the idea of a complete contingency plan like the one the boss left in the sealed envelope. Notice that she did not lose any flexibility or control by going away since the outcome is the same as if she had been there. For simplicity in the following examples we shall use letters to label the outcomes, without mentioning what payoffs may be involved for one player or the other. The game may or may not be zero-sum. For ease of discussion the different alternatives are also labeled.

In Tree 2.9 player A clearly has three possible strategies, *a, b,* or *c*. However, a complete contingency plan for player B is more

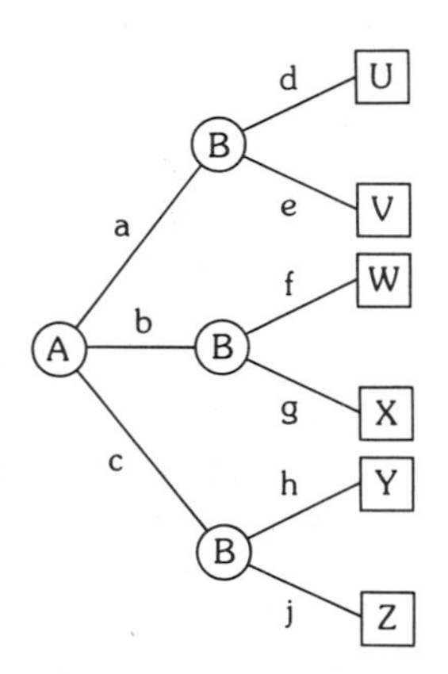

Tree 2.9 A game tree.

complicated because B must state in advance what he will do in response to A and does not know what A will do. Again, think of B as having an assistant who is given instructions for what to do in all possible contingencies. The eight different possible sets of instructions are:

☐ **Set 1** *d-f-h,* that is,
pick *d* if A has chosen *a;* and

pick *f* if A has chosen *b;* and
pick *h* if A has chosen *c*.

- **Set 2** *d-f-j,* defined similarly.
- **Sets 3–8** *d-g-h, d-g-j, e-f-h, e-f-j, e-g-h, and e-g-i,* respectively.

The matrix that corresponds to Tree 2.9 is found, as always, by pairing up the choices of the two players. For example, if A picks strategy *b* and B has left instructions to use the strategy *e-g-h,* the result is that B ends up actually using branch *g* of the tree, so the outcome is X. Proceeding in this way, one obtains Matrix 2.5.

	dfh	**dfj**	**dgh**	**dgj**	**efh**	**efj**	**egh**	**egj**
a	U	U	U	U	V	V	V	V
b	W	W	X	X	W	W	X	X
c	Y	Z	Y	Z	Y	Z	Y	Z

Matrix 2.5 Matrix translation of Tree 2.9.

Any tree can be put into matrix form. This fact will not be proved here, but the general method can perhaps be made clear by considering another example, one in which a player makes more than one move. (For simplicity, we will ignore payoffs since they do not affect these considerations.) In Tree 2.10, if player A picks alter-

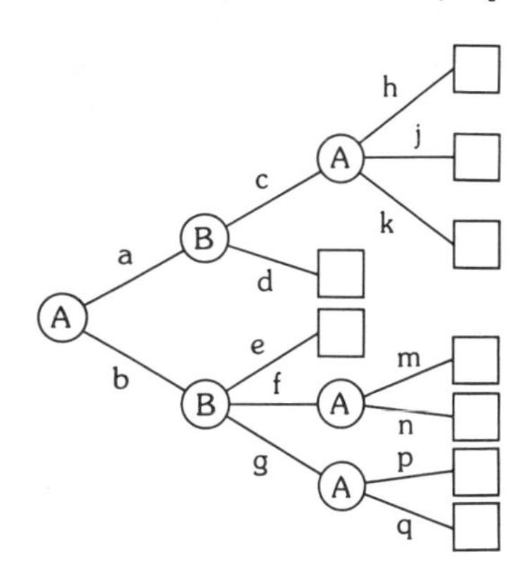

Tree 2.10 A game tree.

native *a* at the start, he may (if B picks *c* in response) subsequently have to decide among *h, j,* and *k*. Thus, for him to write only *a* in advance for his assistant will not provide sufficient guidance. To specify a *strategy* (complete game plan), player A may write *a-h* or *a-j* or *a-k*. On the other hand, A may pick *b* at the start, in which case he must state what he will do in response to both *f* and *g*. Thus *b-n* would *not* be sufficient to specify a strategy since after b is chosen player B may respond with g, in which case A has to have a decision ready for the choice between p and q. On the other hand, b-m-p is a strategy since it clearly tells what to do in all cases. It says: "Pick b first; then in response to f pick m, and in response to g pick p." (If e is chosen, the game terminates.) Note that b-m-p-q and b-m-n are not strategies since each contradicts itself by naming two alternatives at the same choice point. Also a-h-m-p is not a strategy because if a is chosen then m and p will never be possible. In view of all these considerations, player A has seven possible strategies: a-h, a-j, a-k, b-m-p, b-m-q, b-n-p, b-n-q. You should now write down all six possible strategies for B; one of them is d-f.

Exercises

○ **1** Construct the matrix for Game 2.2 "Matching Nickels" with Variety-seeker as the row-chooser. Recall that the numbers in the matrix are, by convention, those of the row-chooser.

○ **2** Rewrite Matrix 2.1 with the following changes. Wherever there is a $+1$ (payoff to row-chooser), replace it by $+1, -1$; wherever there is a -1, replace it by $-1, +1$. Thus each cell will have two numbers: row-chooser's payoff followed by column-chooser's payoff. Within each cell, the numbers add up to 0; you have thus shown the "zero-sum" property explicitly.

○ **3** **(a)** Construct the matrix for "Matching Nickels with Pennies from Heaven," played as follows. Players play "Matching Nickels" and, whatever the outcome, each receives a penny from a mysterious source. Thus a player ends up with either $5 + 1 = 6$¢ or with $-5 + 1 = -4$¢. Make Matchmaker the row-chooser and show both players' payoffs separated by a comma as in exercise 2.

(b) Add up the two numbers within each cell. Notice that the sum is the same in each case. What is that sum? Since these sums are not zero, this is not a zero-sum game; but since they are all the same, it is a constant-sum game.

(c) Constant-sum games are "strategically equivalent" to zero-sum games. To see this, notice that the pennies showered on the players merely make them a little richer. Since the wealth of the players is not included in these game descriptions, this change cannot be significant. Construct another constant-sum matrix and construct a zero-sum matrix to which it is strategically equivalent.

(d) Is Matrix 2.3 zero-sum, constant-sum, or variable-sum? Very briefly, why?

○ **4** Construct the matrix for Game 2.4.

○ **5** Construct the whole tree for Game 2.6.

○ **6** Which player can surely win in Game 2.10? How?

○ **7** **(a)** Draw the tree for Game 2.11.

(b) How can A insure getting at least 1¢?

(c) Can A insure getting 2¢?

(d) Can B be sure of losing no more than 1¢? How?

○ **8** Player I goes first and may choose the letter F or H. Then player II goes and may choose E or U. Then player I goes again and picks R or T. If a word is formed by the three letters in the order chosen, player I wins.

(a) Draw the tree.

(b) Who can guarantee a win? How?

(c) Suppose payoffs are +1 for winning and −1 for losing unless the word FUR forms, in which case the stakes are tripled. What should player I then do?

○ **9** In the game described in exercise 8, suppose that the rules are altered so that if the two-letter word HE forms, the game terminates (with player II winning)

(a) Describe all of player I's strategies. You may use abbreviations, as in Matrix 2.5, if you define them.

(b) Describe all of player II's strategies.

(c) Draw a matrix of the right size and fill in all the entries in one row of it.

○ **10** In the game of Tree 2.10,

(a) Name the six possible strategies for player B.

(b) State a strategy for A and one for B which, if used together, will result in the lowermost outcome box in the tree.

(c) Is there more than one possible answer to part (b)? Explain briefly.

○ **11** Suppose A threatens B as follows: "If you do *x* then I will do *y* so you better do *z*." Assume that B goes first, so A has to make a decision only if the threat goes unheeded. If A does make a decision it is between "y" and "not y" (carry out the threatened action or not).

(a) Make a tree showing players and possible moves. Put the labels P, Q, and R on the three possible outcomes (the "leaves" of the tree).

(b) How does the relative desirability of P, Q, and R to the two players affect the credibility and effectiveness of the threat? (Your answer at this point will be based on judgment and intuition. In Chapter 6, these matters will receive a fuller game-theoretic analysis.)

(c) Discuss similarities and differences among the meanings of "threat," "promise," and "warning" in terms of trees like the one in your answer to part (a).

(d) Look at the description of Game 6.1 in Chapter 6.

There the U.S. threatens South Africa that it (South Africa) had better threaten Rhodesia. In this "threat-to-threaten" situation, which of the three players goes first? Which goes last? Note the order of threatener and threatened player in part (a).

○ **12** Find a newspaper or magazine article with at least two interacting individuals or groups. List all the individuals or groups mentioned and then select those who are key players. Very briefly state why they are important or why others are less important. Your answer should make reference to decisions between (among) alternatives. Often the interesting thing about a news story is not what did happen but why some other thing did not happen. For example, A might have avoided some attractive option, knowing what it would have led B to do. (Note to instructor: Students should submit the article with their answers. Expanded versions of this exercise can be repeated later in the course, when students will be able to use a wider variety of game-theoretic tools in their answers.)

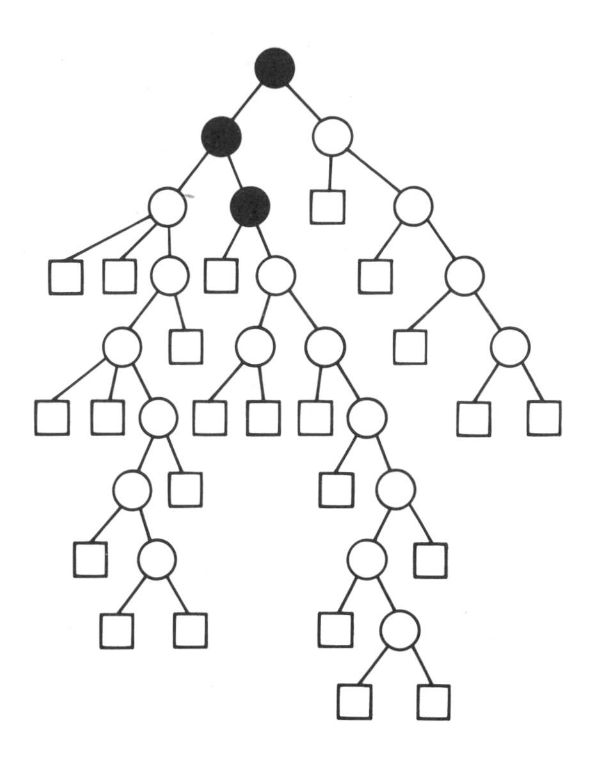

3 Solutions: Two-Person, Zero-Sum Games

If life is at all like a game, then perhaps a good question is, "How can I win?" Or at least, "How can I avoid losing?" Or more generally, "What is the best way to play considering the cards I have been dealt (the situation in which I find myself)?" If life as a whole is too complicated, then perhaps certain aspects or parts of living can be subjected to these questions. Thus, one might seek the ideal strategy for getting a raise, helping to win an election, being widely respected, etc. In this chapter we shall "solve" the problem of what strategy to choose for a certain class of games.

Although good strategies are what we seek, this book is *not* a handy pocket guide to action in interactive-decision situations.

Rather, as the book's title suggests, the purpose is to foster insight into social phenomena through a kind of model called games. This use of games as models is amplified in Section 3.1 with particular reference to the notion of "solutions," the main topic of this chapter.

Utility theory, sketched in Section 3.2, has to do with people's preferences and is an important underpinning for game theory. It is the study of how we determine what particular values ought to be used as payoffs in a game matrix or tree. These values must reflect the players' preferences in a meaningful way or else the whole strategic analysis is undermined.

The notion of "rationality" in interactive-decision-making is introduced in Section 3.3. It turns out to be possible, though not simple, to find a concept of rational solution, i.e., a sensible and consistent set of principles on which to base decisions. These principles are valid for two-person, zero-sum games, but will lead to profound and interesting problems in later chapters.

Even in the restricted domain of two-person, zero-sum games, the idea of solution is sufficiently complex that its correctness is not obvious. Therefore an informal proof is presented in Section 3.4 for the simplest case, the one in which each player has two alternatives to choose from. The last section deals with more complex cases. (The less mathematically oriented reader may wish to skip these sections, at least on the first reading.)

3.1 The Concept of Solution

Very roughly speaking, a solution to a zero-sum game consists of a statement of what a player should do and what he can expect to get by doing it. It is important to notice that this use of the word "solution" is only used with respect to precisely formulated *games*. We do not speak of solutions to *situations* or claim to solve the problems of life. Nevertheless, we do claim to gain insight. Let us see how this may be achieved.

The first step is to take a real-life situation and try to simplify it in a reasonable way, extracting what seem to be its principal features, to form a model of the situation. If that model is a type of game for which we have solution techniques, then the next step is to solve our particular game. By translating back from the game into the real-life situation, we may hope to find a "solution to the situation," a prescription for how to decide. However, any such prescription will only be as good as the model. If we have done a good job of making reasonable assumptions, then our solution may translate into a reasonable guide to action, but if our model is poor, then the solution may be worthless.

In particular, there is a very severe limitation on the usefulness of the results in this particular chapter. This chapter deals with zero-sum games and, as pointed out at some length in Section 1.1, real-life interactive-decision structures are often variable-sum. A zero-sum model used with a variable-sum situation cannot hope to provide much insight or guidance.

Even though the results in this chapter are of limited applicability, they provide important perspective on later material. Moreover, the *concepts* introduced here are crucial. This treatment of zero-sum games includes several key definitions and principles that will be used in the analysis of variable-sum games in the following chapters.

A solution to a two-person, zero-sum game means answering two questions: "What is the best thing to do?" And, "What payoff will be the result?" The difficulty is that the second question depends on the other player's choice, and the first question depends on the answer to the second (what is the best choice depends on what it gets us, compared to other alternatives).

These questions can also be asked (cautiously) of real-life situations. Take elections for instance. It is not possible for both (all) candidates to win. Thus, if each side asks the game theoretician for a winning strategy, she cannot possibly provide one for each of them. Nor can she possibly provide each side with a strategy for surely avoiding losing. Thus, the most we can ask her for is advice as to how to do as well as possible under the circumstances. "As well as possible" may turn out to be a loss, as for example in the game of "heads I win, tails you lose," a game so unfair that not even a game theoretician can help you avoid losing.

Even if we set our sights realistically, however, and merely ask to do as well as possible, difficulties remain. For example, consider Game 2.1, "Matching Pennies." What is the best strategy? Doing "as well as possible" seems to mean getting +1, but there is no way to be sure of getting it unless you know what the other player will do. And, of course, if the other player knew what *you* would do, then you would be stuck with −1. But neither one of you knows what the other will do, so neither strategy (heads or tails) seems better than the other and neither payoff (+1 or −1) seems appropriate to be called "doing as well as possible."

In the face of these perplexities, game theory prescribes a randomized strategy. In "Matching Pennies," for example, the solution is to base your decision on the flip of a coin so that you come out even, "on average." For more complex situations one must use more complex random choices, as we shall see in Sections 3.4 and 3.5. Game theory succeeds in providing solutions to all two-person, zero-sum games.

3.2 Preferences

Finding a strategy for doing as well as possible under the circumstances was the problem posed in the preceding section, and we shall answer this problem for zero-sum games in later sections. However, "doing well" means different things to different people, so we must first discuss the notion of individual preference. Some people may not always want to win or get large amounts of money at the expense of others. "Time is money," it is said, and different people value their time differently. What is time lost for one person may be a good time to another. Loss or gain of affection, power, or respect may be worth more than many dollars or hours and may not be purchasable at any price. Even for a person whose main interest is money, his first million may be more important than his second. For all these reasons we measure success not in dollars or in victories but on a *utility scale.*

An understanding of utility scales is crucial if the solutions obtained later in this chapter are to have any meaning. There are several kinds of utility scales, but we shall introduce only the ordinal scale and the interval scale. These two scales make certain assumptions about what people can reliably tell you about their preferences.

3.2.1 The Ordinal Scale

To say that a person's preferences can be measured on an ordinal scale simply means that they are *in an order.* In other words, the various possible outcomes of a situation can be laid out along a line in such a way that for any two outcomes the preferred one lies to the left of the other. Does this procedure really work for real-life situations? That is, are preferences really like left-right comparisons?

Suppose that Laurent is a participant in a game or situation with possible outcomes A, B, and C, and he tells us that he prefers A to B and B to C. No comparison has yet been made between A and C. Let us tentatively represent Laurent's preferences by placing A to the left of B and B to the left of C. Notice that when we do this, A and C are automatically placed in a left-right relationship to each other, even though we have not yet asked Laurent which of them he prefers. It might seem "logical" that A *must* be preferred to C, but that is Laurent's choice to make. We are social scientists engaged in finding out what people believe, not telling them what they ought to believe in accordance with what seems logical to us.

So far we may seem to have said very little that is not obvious, but surprisingly weighty concepts are involved here so we had better step back a moment and see just what is implied by what we are doing. Points lying in a line, and their left-right arrangement, taken together, are an example of a *model.* On the one hand, we must be careful to keep the "model world" separate from the real world—they are two different things. On the other hand, the whole purpose of the model world is that it serves as an analogy to the real world. If it is a good model, then true statements in the model world will also be true in the real world. The model world of points lined up left to right is characterized by the following properties:

☐ **1 Comparability** For any two points we can tell which one lies to the left.

☐ **2 Transitivity** If A is to the left of B and B is to the left of C, then A is to the left of C.

Having characterized the model world, let us get back to the real world of people and their preferences. In particular, we now ask whether people's preferences satisfy the properties of comparability and transitivity. If they do, then we are justified in representing Laurent's preferences by points in a row. Stated in terms of preferences, 1 and 2 become 1′ and 2′:

☐ **1′ Comparability** For any two outcomes a person prefers one to the other.

☐ **2′ Transitivity** If someone prefers A to B, and B to C, then he prefers A to C.

Violation of 1′ would occur if we asked, "Which do you like better, milk or coffee?" and our respondent said, "Coffee on a cold day, milk on a hot day, since the heat of the day determines the heat of my beverage. I can compare coffee to tea or milk to orange juice, but I cannot compare coffee and milk to each other in general."

Violation of 2′ is also possible. Suppose someone prefers an office on the tenth floor to one on the ground floor because of the view, and that, on the same basis, the twentieth floor is preferred to the tenth. Now we ask this individual for a preference between the ground floor and the twentieth and, lo and behold, she prefers the ground floor! Why? Because she does not want to waste so much time in the elevator—a consideration that suddenly becomes very important when considering two floors so far apart.

Thus it is at least conceivable that 1′ or 2′ could be violated for some people, for some of their preferences. In fact, however, we find few violations; most preferences of most people satisfy conditions 1′

and 2′. This fact is a discovery about the world, obtained by doing an experiment, namely asking people what they prefer. The experimental results tell us that the model is appropriate, that the real world conforms quite well to the model. This particular model is called a *simple ordering,* and we say that preferences are *simply ordered.* This is equivalent to saying that preferences can be measured on an ordinal scale.

3.2.2 The Interval Scale

The interval scale involves stronger claims than the ordinal scale. To say that an individual's preferences fit an interval scale means not only that those preferences satisfy an ordinal scale but also that the individual can give consistent statements of preference between probabilistic lotteries. Judgments about lotteries, to be discussed shortly, will turn out to imply judgments about *how much* one thing is preferred to another. An ordinal scale can represent the notion that one thing is preferred to another but not that one thing is preferred by a wide or narrow margin. The interval scale, in contrast, puts an exact numerical value on the amount by which one outcome is preferred to another.

Numbers may seem to provide an unfeeling medium for discussing real people making real decisions. However, the numbers on an interval scale facilitate comparison of diverse situations by reducing money, time, health, pleasure, etc., to a common scale of measurement. One may balk at comparing pain with money or death with time, but life presents inescapable decisions involving just such comparisons. The person who fails to buckle up a seat belt is expressing the belief that a small probability of death (the probability that a crash will occur in which a seat belt would make a difference between life and death) is less consequential than the time it takes to put on the belt. The congressman who votes on mandatory safety devices makes a value judgment on lives other than his own.

Suppose that for some student outcome A is the event that he gets all A's and a case of mononucleosis, while outcome B consists of getting all B's and a case of beer. Finally, with outcome C he simply gets all C's. Let us assume, not unreasonably, that A is barely preferred to B, while B is clearly preferred to C. Notice the words "barely" and "clearly." If you did not balk at their use, then you have already realized that it is possible to compare not just the outcomes but *differences between* outcomes. Let a, b, and c be the numerical utility values for A, B, and C, respectively, then $a - b$ (the bare difference) should be less than $b - c$ (the clear difference: $a - b < b - c$. Rearranging this inequality, we get $a + c < 2b$, which yields, upon dividing by 2: $1/2a + 1/2c < b$. This last mathematical statement may be stated in words as "B is preferred to half of outcome A and half of outcome C." Now you cannot half have a case of mononucleosis and half not have it, or, to take another possible pair of outcomes, you cannot be half pregnant and half not pregnant. But we can interpret $1/2a + 1/2c$ as the utility of a lottery ticket that entitles us to get either A or C (but never both), each with probability 1/2. We will assume that players really do have consistent preferences among lotteries of this sort. This assumption plus a few others (see Luce and Raiffa, 1957, for a fuller treatment) enable us to have utility scales on which differences such as $a - b$ and $b - c$ are meaningful. Such scales are called *interval scales* of utility. ("Difference" and "interval" have almost the same meaning).

Our use of lotteries and interval scales will demand some understanding of probability, in particular the notion of *expectation.* (The reader with a knowledge of probability may wish to skip the next few paragraphs.) Suppose a fair coin is flipped and you are to win 12¢ if it comes up heads but only 6¢ if it is tails. The expectation of your winnings in a case like this, where both (or all) outcomes are equally likely, is simply the average value of the possible outcomes. In this case it is $(6 + 12)/2 = 9$¢. One way to think about what expectation means is to ask how desirable it would be to find yourself in this situation. Clearly, it is worth more than 6¢ but less than 12¢, perhaps 9¢. From another point of view, suppose that this activity were repeated many times. Then you would probably get 12¢-wins about half the time and 6¢-wins about half the time and so come up with about 9¢ per toss, on average.

Suppose that the coin in the preceding example is biased toward heads and that the probability of heads is 2/3. This new situation is more valuable to you since you are more likely to get the larger amount, 12¢. This improvement is reflected in an increase in the expectation of your winnings, which is now $(2/3)(12) + (1/3)(6) = 8 + 2 = 10$¢. Expectation is defined as a weighted average of outcomes where the weights are probabilities. This definition could also have been used in the equal-probabilities example above to get $(1/2)(6) + (1/2)(12) = 9$¢.

Although we have interpreted expectation in terms of how "desirable" the situation is or in terms of a long-run average when the situation is repeated, it is important to note that the *definition* of expectation does not rely on these interpretations. Expectation is found, as shown in the examples, by multiplying each outcome by its probability of occurrence and adding these products.

Lotteries need not be restricted to just two outcomes. The above definition of expectation as a weighted average can be used with any finite number of outcomes. For example, suppose a fair die is to be rolled and you win 6¢ if it comes up a 6, 3¢ if it is a 3, and otherwise lose a penny. The expectation of your winnings is then

$$(1/6)(6) + (1/6)(3) + (4/6)(-1) = 5/6¢$$

The above examples have been couched in terms of money, but the concept of expectation can be applied equally well to utility. Suppose that Harmon is deciding whether to go to the beach. If he goes and it stays sunny, he will have a good time; but if it rains, he will have wasted the day. If he stays home, he will have a so-so day, rain or shine. The probability of rain is 1/4.

To be more precise about Harmon's value-system, let us assume that "good time," "so-so day," and "wasted day" can be appropriately replaced by interval-scale utility values of 8, 3, and 0, respectively. Staying home is a sure 3, while going to the beach is equivalent to taking a lottery ticket that yields 8 with probability 3/4 and 0 with probability 1/4. The expectation of the utility of such a ticket, or, briefly, its *expected utility,* is $(3/4)(8) + (1/4)(0) = 6$. Since 6 is higher than 3, we may anticipate that Harmon will make the trip. In fact, if Harmon does *not* go to the beach, then either his utilities are not 8, 3, and 0 or else his estimate of the probability of rain is not 1/4.

Suppose we are told that on Emily's utility scale the outcomes J, K, and L have utilities of 4, 5, and 7, respectively. From this we know that she must have preferences as follows. She prefers L to K, she prefers K to J, and she is indifferent (has no preference) between K and a lottery in which she gets J with probability 2/3 and L with probability 1/3. These probability values have been obtained from the following calculation. Let x be the probability of L in a lottery between J and L that has exactly the same value to Emily as surely getting K. Then,

$$\begin{pmatrix}\text{utility}\\ \text{of J}\end{pmatrix}\begin{pmatrix}\text{probability}\\ \text{of J}\end{pmatrix} + \begin{pmatrix}\text{utility}\\ \text{of L}\end{pmatrix}\begin{pmatrix}\text{probability}\\ \text{of L}\end{pmatrix} = \text{utility of K}$$

$$(4)(1 - x) + (7)(x) = 5$$

$$7x - 4x = 5 - 4$$

$$x = \frac{1}{3}$$

Thus, Emily is indifferent between K and a lottery in which J has probability 2/3 and L has probability 1/3.

It is important to realize that if another player, Emmett, has utilities of 40, 50, and 70 for the outcomes J, K, and L, respectively, then all the things said about Emily are true for Emmett as well. Not only is the order of his preferences the same, but again 1/3 is the probability that must be assigned to L in a lottery between J and L to make the lottery exactly as valuable to Emmett as K. The same is true of Eliza, whose respective utilities are 340, 350, and 370, and of Edmund with −11, −10, and −8, as the reader may check using the formula

$$\frac{2}{3}\,(\text{utility of J}) + \frac{1}{3}\,(\text{utility of L}) = \text{utility of K}$$

Scale changes of this sort are just like those involved in converting temperature from Fahrenheit to centigrade (Celsius) or vice versa.

Returning to Emily, suppose that all we knew about her utilities was that J and L had the values 4 and 7, respectively, and that K was at some intermediate point. We could then try to find the exact value of K by asking Emily to compare K to lotteries between J and L with various probabilities. If it turned out she preferred K to any J-and-L lottery for which the probability of L was below 1/3, and if moreover she preferred a J-and-L lottery to K whenever the probability of L was above 1/3, and if finally she was indifferent between K and a J-and-L lottery in which the probability of L was exactly 1/3, then we would confidently assign the number 5 as the utility of K to Emily.

Just as a player's preferences must be transitive for us to use an ordinal scale of utility (Section 3.2.1), so must certain requirements be met with the interval scale. In addition to properties 1′ and 2′ in Section 3.2.1, the interval scale requires a third property: consistency of expected utilities.

The idea of this property can be seen by continuing the last example. Suppose that outcome Q is assigned a utility in the same way that K was, namely by having Emily compare it to various lotteries of J and L. Further, suppose that Emily is indifferent between Q and a lottery in which J has probability 1/3 and L has probability 2/3; Q is thus assigned the value (1/3)(4) + (2/3)(7) = 6. It makes sense at this point to anticipate that Emily prefers Q (with utility 6) to K (with utility 5), and let us imagine that she indeed does have such a preference. But something more precise now also follows from these interval-scale values: Emily must be indifferent between Q (=6) and a lottery between K (=5) and L (=7) in which the two outcomes each have probability 1/2.

We cannot tell Emily to be indifferent with exactly this probability in the lottery, but unless she is, her utilities cannot be represented by an interval scale. In reality, people's preferences among lotteries are *not* always exactly consistent in this way. Nevertheless, they are usually close enough that the interval scale is a reasonable model of preferences, and we shall use it henceforward.*

3.3 Principles of Maximizing

Now that utility scales have been defined, it is possible to charge ahead with our original problem of "doing as well as possible." Since *by definition* any player prefers a higher utility, we can simply replace "doing as well as possible" by "maximizing one's utility." It is, after all, only by virtue of someone telling us that he prefers A to B that we assign A a higher utility than B for him. By this definition, even an altruist seeks higher utility, since his concern for others is incorporated into his stated preferences.

It turns out that we can say quite a lot about how a player can maximize his utility in a game without making any use of the interval scale. In fact, the two principles of maximizing in this section depend only on the ordinal scale, so that all we shall speak about here is whether one outcome is preferred to another. Lotteries are not mentioned, nor is one outcome ever said to be "much better" or "barely better" than another. Although numbers will be used in the matrices, we will make use only of the relative sizes of these numbers. That is, we will not use any interval-scale information, only ordinal-scale information. Quite the opposite will be the case in the following section (3.4), where the interval scale will be crucial.

What reasoning can players use to "do as well as possible"? Consider, for example, the zero-sum game of Matrix 3.1. One of A's

	B	
A	4	3
	0	−2

Matrix 3.1 Game with dominant strategies for both players. The top row is a dominant strategy for player A since 4 exceeds 0 and 3 exceeds −2. The right column is a dominant strategy for B since −3 exceeds −4 and 2 exceeds 0 (player B's payoffs are the negative of those for A).

* Important recent work has uncovered certain systematic deviations of people's lottery preferences from the inverval-scale requirements. "In particular, people underweight outcomes that are merely probable in comparison with outcomes that are obtained with certainty" (Kahnemann and Tversky, in press). Even more fundamental difficulties with preference are summarized by Grether and Plott (in press).

choices is clearly superior to the other, no matter what B does. If A thinks B will pick the left column, then she (A) will be choosing between 4 and 0, so she will pick the top row to get 4. On the other hand, if she thinks B will pick the right column, then her choice is between 3 and −2, and again the top row yields a higher payoff. Thus, no matter what A may think B will do, she is always better off with the top row. In such a case we say that the top row *dominates* the bottom row. Since A's strategy of picking the top row dominates *all* her other strategies (there is only one other in this case), it is called the *dominant strategy* for player A.

In this same game one of B's strategies dominates the other. Recall that the payoffs stated in the matrix are the negatives of B's payoffs, so, looking at the game a little differently, B wants to *minimize* the payoff to A, since in this zero-sum game A's gain is his loss and vice versa. More precisely, to say that this matrix is zero-sum is to state in effect that player B's preferences are accurately represented by utility values of −4, −3, 0, and 2, so that, for example, the lower-right cell is the best for B.* If B thinks A will pick the top row, then he (B) can get either −4 or −3; since by definition of utility he prefers −3, he will choose the right column. Also, if A were to pick the bottom row, B would do better with the right column since he prefers +2 to 0. So B's dominant strategy is the right column.

As a final comment on Matrix 3.1, recall that a solution involves not only a choice of strategy but also a statement of results. If both players use their dominant strategies in this game, the resulting payoff will be 3 (+3 for A and −3 for B), so we say that the *value* of the game is 3, meaning that this is our anticipation of what the payoff will be.

The analysis of the preceding game can be summarized in the form of a guideline for decision-makers:

◀ **Rationality Principle 1**

If there is a dominant strategy available to you, use it.

This principle of maximizing hardly needs justification, once one understands what "dominant" means. The word "rationality" is a vague and loaded word, but we are trying to make it mean something precise and see where the attempt leads us. This guideline may seem so simple that indeed any "rational" or reasoning person

* Since payoffs are on an ordinal scale, all this means is that B's preference order is the reverse of A's. This reversal is, however, the main idea of the zero-sum game, as discussed in Section 1.1.

ought to heed it. For the purposes of this chapter the reader is encouraged to believe that there really can be behavior guidelines so reasonable that any reasonable person would agree with them. This paves the way for a deeper understanding of the "dilemma games," discussed at length in Chapters 4 and 7, where this view of rationality is seriously undermined.

Use of the expression "rationality principle" also suggests that we are prescribing behavior, telling people what to do. The social scientist seeks to *describe* rather than *prescribe,* that is, to observe and ultimately to explain. Therefore, we should remember that it is a question to be decided by many observations whether or not people's decisions really do conform to our strategic analysis.

Now let us examine Matrix 3.2 for dominant strategies. Player A does not have a dominant strategy, but player B does. If A thought

	B	
A	−2	1
	−3	4

Matrix 3.2 Game with dominant strategy for only one player. Since 2 exceeds −1 and 3 exceeds −4, the left column is a dominant strategy for player B. There is no dominant strategy for A. If A assumes that B will use his dominant strategy, then A picks the top row to get −2, not −3.

B would pick the left column, she would take the *top* row to get −2, not −3. But if B were to pick the right column, A would want 4, not 1, which is in the bottom row. Thus, although A has no dominant strategy, she nevertheless has a sound basis for making a choice.

Suppose that A puts herself in B's shoes. Clearly, B has a dominant strategy because his best two payoffs are both in the left column (2 and 3). So A can be pretty sure that B will pick the left column, leaving A with a choice between −2 and −3, from which A can reasonably decide to go for the −2 by picking the top row.

◀ **Rationality Principle 2**

Assume that the other player is "rational" and respond optimally to what he will do.

To "be rational" in this sense is to act according to whatever principles of rationality we happen to be talking about. In the example just discussed, player A uses Rationality Principles 1 and 2 together to settle on the top row. Notice that in Matrix 3.1 players can base sound decisions directly on their own preferences. In contrast, Matrix

3.2 requires player A to think "interactively," to look at the game from the other player's viewpoint.

The notion of an *equilibrium outcome* will turn out to be important. For an outcome to be an equilibrium, it must be the case that if any one player had chosen differently, he would have done worse. For example, in Matrix 3.2 the upper-left cell is an equilibrium because if A alone had chosen differently, he would have gotten −3 (lower than −2), while if B alone had chosen differently, he would have gotten −1 (lower than +2). Put differently, there is no "regret" when an equilibrium occurs, since even if a player had known what the other was planning to do, he still would have made the choice he did.

An equilibrium may exist even in the absence of dominant strategies. Consider, for example, Matrix 3.3. First, confirm that

B

	5	0	−3
A	3	2	4
	−2	1	6

Matrix 3.3 Game with a stable outcome. Neither player has a dominant strategy (you should confirm this). At the center cell player A gets 2 while B gets −2. If A alters his choice while B sticks with the center column, then A gets 0 or 1, both less than 2. If B alters his choice while A sticks with the center row, then B gets −3 or −4, both less than −2. The center cell is therefore an *equilibrium*.

neither player has a dominant strategy by checking all possible pairs of alternatives for each player. (Thus, for A compare the top and middle rows, the top and bottom, and the middle and bottom.) Next, check individual cells for equilibrium. (Notice that a *strategy* may be dominant, while a cell or *outcome* may be an equilibrium.) The upper-left cell is not an equilibrium since B would "regret" this outcome in the above sense. Either of the other columns would be better for B (either 0 or 3 is better than −5). The bottom-center cell would be regretted by both players. For example, A could have gotten 2 instead of 1 by selecting the middle instead of the bottom row. Check all the remaining cells for equilibrium (the correct result is stated in the caption of the matrix).

There are also games that have neither a dominant strategy nor an equilibrium. A simple example is "Matching Pennies," as you can verify from Matrix 2.1. How shall a player decide what to do in such a circumstance? We shall consider several possible selection principles, none of which is as convincing as dominance but which at least have the virtue of always giving us a decision.

Consider Matrix 3.4. Imagine that you are player A and that player B is Nature, who makes her choice of columns in an un-

	B			
A	4	3	1	0
	2	2	2	2
	3	3	3	0
	3	2	2	1

Matrix 3.4 Different criteria specify different rows.

fathomable, nonstrategic way. Nature does not receive a negative payoff when you get a positive one and does not care what happens. Her decision is arbitrary. The player might be a farmer choosing a planting strategy and the columns different kinds of weather. What should you (A) do?

One possible approach is to find the highest possible outcome and try to get it, by picking the row in which it occurs. In Matrix 3.4 this means picking the top row. The rationale for this selection principle of "go for the best" is that since anything can happen, you might as well not deny yourself your best possibility. After all, you might get it. This is an optimistic principle.

On the other hand, a pessimistic soul would want to avoid the worst payoffs. In Matrix 3.4 this would mean eliminating from consideration the first and third rows (which contain 0). Pursuing this line of reasoning, note that the lowest payoff in the second row is 2, while in the bottom row the lowest is 1. Therefore, the second row is the safest—it guarantees a minimum payoff of 2.

This last approach has a special, widely used name, worth remembering and understanding. It is called the *maximin principle* and works as follows. First find the smallest payoff in each row; in Matrix 3.4 these are 0, 2, 0, and 1 in order from top to bottom. These numbers are called the row minima (plural of "minimum"). Next, pick the largest, or maximum, of these. This number is the maximum of the minima, or "maximin." The *maximin payoff* in Matrix 3.4 is 2, which is in the second row, and that row is therefore called the *maximin strategy*.

The optimistic, go-for-the-best approach can be described in a similar manner and is called the *maximax principle.* First find the largest (maximum) payoff in each row (4, 2, 3, and 3 in Matrix 3.4).

Then pick the largest (maximum) of these (4 in our example). This number is the maximum of the maxima, or "maximax." The maximax principle is to pick the strategy containing the highest or *maximax payoff*.

Another possible strategy is the *maximum-average principle.* First find the average for each row. In Matrix 3.4 the top row has an average of $1/4(4 + 3 + 1 + 0) = 2$. The third row has the highest average, 2 1/4, and so would be selected. This principle would make sense as a strategy if you knew that a column would be chosen randomly with all columns equally probable. The principle is neither optimistic nor pessimistic, nor is it particularly well justified since we have assumed that you do not know how the column is to be chosen.

The bottom row of Matrix 3.4 has the peculiar property that if you (player A) pick it, you can guarantee yourself never to be very "regretful." In any column the payoff in the bottom row is within one point of the highest possible for that column. Such a claim cannot be made for any other row. This strategy is therefore the one selected by the *minimax-regret principle*.

The four selection principles just described are similar in one important way: *any one of them will pick the dominant strategy when one is available.* (You should check an example to see that this is true in that example and then try to state the reasons why it is always true.) Despite this similarity, we have seen that all of these principles are logically distinct since they all may lead to different choices in a matrix without a dominant strategy. For example, in Matrix 3.4 a maximax player will pick the first row, a maximin player the second, a maximum-average player the third, and a minimax-regret player the fourth.

3.4 Probabilistic Strategies and a Solution

A two-person game in which one player has m alternatives and the other has n is called an $m \times n$ ("m by n") game. (By convention, the first number is always the number of rows, the second is columns.) The smallest possible size of a game (matrix) is 2×2, since if a player had only one alternative, there would be no real decision to make. This section shows how to find solutions for all 2×2 zero-sum games.

First, the structure of a game is not changed by interchanging its rows. Such a change merely amounts to renaming the alternatives: the alternative previously called "top row" is now called "bottom row'' and vice versa. The columns can also be interchanged. Using these operations, we first get the highest payoff (to A) into the upper-left corner. For example, Matrix 3.5a can be rewritten as Matrix 3.5b by interchanging the rows. Interchanging the columns yields Matrix 3.5c. All three matrices are strategically equivalent.

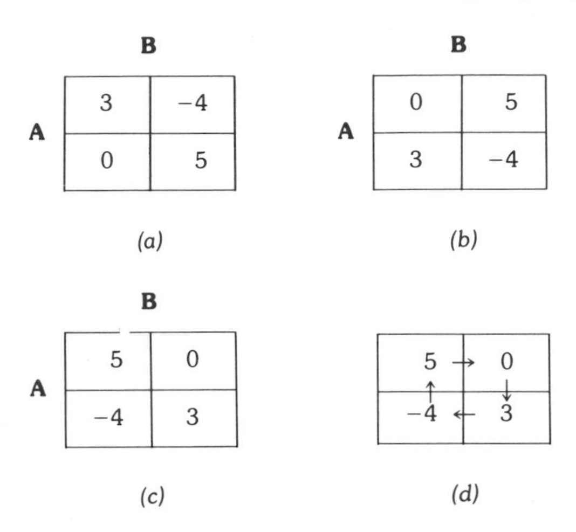

Matrix 3.5 Matrix (*a*) is rewritten into (*b*) by interchanging the rows, without changing the structure of the game. Matrix (*b*) is rewritten into (*c*) by interchanging the columns. Matrix (*d*) shows the cycle of changes of choice for Matrix (*c*).

Next we turn our attention to the second-highest payoff. There are three places where it can now be: (1) upper right; (2) lower left; or (3) lower right. In case 1 the two highest payoffs would be in the top row, making it a dominant strategy for A. In case 2 the two highest payoffs would be in the left column, making the *right* column dominant for B. These first two cases have already been provided with solutions in Section 3.3. Finally, in case 3 the two highest payoffs are diagonally placed and neither player has a dominant strategy. This is the case for Matrix 3.5c, as it is for ''Matching Pennies.''

Such a game not only lacks dominant strategies, it also does not have an equilibrium. For example, consider Matrix 3.5c. The upper-left cell is not an equilibrium, since if B thought A was going to use the top row, then B would shift to the right column, since he prefers 0 to -5. But then if A thought B were going to do that, he in turn would use the bottom row, getting 3 instead of 0. And so the reasoning continues, following the arrows in Matrix 3.5d, which has the same numbers as 3.5c. Thus, this game has no stable outcome.

Notice that the cycle in this example is no accident. Any 2×2 zero-sum game in which the two largest payoffs are diagonally opposite each other will have a cycle and hence no stable outcome.

The reader can check this by making a 2×2 matrix with Hs (for "high" payoff to A) on one diagonal and Ls (for "low" payoff to A) on the other and then drawing arrows as in Matrix 3.5d.

What this means is that *no* selection principle that prescribes a single alternative for a player can possibly work properly in such games. Suppose, for example, the maximax selection principle is rational. Since A is rational, he chooses his maximax alternative, i.e., the top row, which contains 5, his best possible payoff. But because maximax is the rational way to behave, B knows A will do this and so will pick the second column, getting 0 instead of −5. Notice that B is *not* following maximax, since his best possibility is +4 in the *left* column. Thus, if one were to assume that the following principles together constituted "rationality," then player B would have a nervous breakdown (Principle 2 was introduced earlier):

◀ **Rationality Principle 2**
Assume that the other player is "rational" and respond optimally to what he will do.

◀ **Rationality Principle 3**
Always follow maximax.

Principle 3 by itself counsels the left column, but the two principles together counsel the right column (in anticipation of a maximax choice by the other player).

How can this kind of game be solved? Can we find a set of "rationality principles" that are consistent in that they do not lead to conflicting advice? If one imagines that a game will be played many times, then it might be possible to work out a method of switching back and forth, so the other player is kept guessing. We shall take such an approach, and then discuss whether the results obtained are also relevant if the game is to be played only once.

Forgetting about the general problem of finding a selection principle good for all 2×2 zero-sum games, let us simply focus on the case at hand, Matrix 3.5c. Since we are imagining the game to be played over and over, maximax for A now means choosing the top row all the time. But A would be foolish to do this since B would figure it out and respond by always picking the right column for a payoff to each of 0. We shall soon see that A can do better than this, but first note that for A to pick the other (bottom) row repeatedly is even more foolish.

Now suppose A decides to pick one row some of the time and the other some of the time. If he does so in a regular pattern, *any* regular pattern, sooner or later B may figure it out and anticipate A's choice at every play of the game, keeping all payoffs at 0 or −4.

It is possible, however, for A to switch back and forth without following a pattern, that is, by choosing randomly. Then B cannot possibly discern a pattern because there is none to discern. For example, suppose A flips a true coin each time to determine his choice. Then whenever B picks, say, the left column, the payoff to A is equally likely to be 5 or -4, i.e., each of these payoffs has probability 1/2. The expectation of the payoff, or simply *expected payoff,* is found according to the definition in Section 3.2.2; in this case it is

$$\frac{1}{2}(5) + \frac{1}{2}(-4) = 2\frac{1}{2} - 2 = \frac{1}{2}$$

Whenever B picks the right column, the payoff to A is equally likely to be 3 or 0 (each has probability 1/2), so in this case A's expected payoff on average is

$$\frac{1}{2}(3) + \frac{1}{2}(0) = 1\frac{1}{2}$$

Thus, against a coin-flipping A, player B should always use the right column, thereby minimizing his (B's) losses by keeping his payoff at $-1/2$ on average rather than $-1\ 1/2$. A coin-flipping player A can therefore anticipate a payoff of 1/2 in the long run.

Player A can do even better in this game by flipping a biased coin so that he picks the top row with probability 7/12. (We will see below why the probability 7/12 has a special significance in this example.) In this case, when B uses the left column the expected payoff to A is

$$\left(\frac{7}{12}\right)(5) + \left(\frac{5}{12}\right)(-4) = 1\frac{1}{4}$$

while when B uses the right column, the expected payoff is

$$\left(\frac{7}{12}\right)(0) + \left(\frac{5}{12}\right)(3) = 1\frac{1}{4}$$

In this way A can arrange to have an expected payoff of 1 1/4 no matter what B may do. Player A can even notify B of his plans and tell him the exact probabilities to be used. There is no way for B to use this information to alter the expected payoff.

At this point some simple definitions will be useful.

☐ **1 Pure strategy** A single alternative, i.e., any single row or column.

☐ **2 Mixed strategy** A probabilistic mixture of a player's alternatives. For example, player A above uses a (7/12, 5/12) strategy.

☐ **3 Maximin pure strategy** Among pure strategies *only,* the one that is maximin. (This is our old definition, but now we are explicitly excluding mixed strategies.)

☐ **4 Maximin mixed strategy** Among *all* strategies, pure or mixed, the one that is maximin.

Notice that the maximin mixed strategy is defined as the maximin among *all* strategies. This is because pure strategies can be regarded as a special case of mixed strategies. For example, (0, 1) for player A means that the probability of the top row is 0, i.e., it is just not used, while the bottom row is sure to be used since its probability is 1. Therefore, (0, 1) is a pure strategy, but technically it has the form of a mixed strategy.

To determine the maximin mixed strategy for player A in Matrix 3.5c, we examine the effects of various mixed strategies that player A might use. A few mixed strategies are shown in Matrix 3.6,

			Minimum in row
	5	0	0
	−4	3	−4
(1/2, 1/2)	1/2	1 1/2	1/2
(7/12, 5/12)	1 1/4	1 1/4	1 1/4

Matrix 3.6 An enlargement of Matrix 3.5*c*, showing two possible mixed strategies.

which includes the pure strategies of Matrix 3.5c as its first two rows. In the new, larger matrix the last row has the largest minimum, and therefore (7/12, 5/12) is the maximin among the mixed strategies that we happened to include in this matrix.

More importantly, no matter what other mixed strategies are considered, 1 1/4 is the best that player A can guarantee himself. For if the probability of the top row is increased beyond 7/12, say to $7/12 + e$, where e is a very small positive amount, then B can use the right column to make the payoff

$$\left(\frac{7}{12} + e\right)(0) + \left(\frac{5}{12} - e\right)(3) = 1\frac{1}{4} - 3e$$

which is less than 1 1/4. On the other hand, if the probability of the top row is decreased by a small amount e to make it $7/12 - e$, then B can pick the left column and the payoff will be

$$\left(\frac{7}{12} - e\right)(5) + \left(\frac{5}{12} + e\right)(-4) = 1\,\frac{1}{4} - 9e$$

In either event A can be forced to settle for less than 1 1/4 if he deviates from (7/12, 5/12). That mixture is therefore his maximin mixed strategy.

The number 1 1/4 also has a special name; it is called the *value of the game* (to player A). Player A can guarantee that his expected payoff will be this amount but cannot guarantee more. We have therefore found a "solution" to this game from A's point of view, according to the definition of solution at the beginning of Section 3.1, namely a statement of what a player should do and what it gets him.

Now look at the game from player B's viewpoint. He wants to keep the payoff to A as low as possible. What is the best he can do? In Matrix 3.5c if player B uses each column half the time, then player A can use the top row and get 2 1/2. But suppose B uses mixture (1/4, 3/4), i.e., the left column with probability of 1/4 and the right column 3/4 of the time. Then A can use the top row and get 1/4(5) + 3/4(0) = 1 1/4 or use the bottom row and get 1/4(−4) + 3/4(3) = 1 1/4. Thus B can guarantee that the expected payoff to A will be no more than 1 1/4, the same value that A could guarantee as his own minimum. This number, the value of the game (1 1/4 in this example), can thus be guaranteed by either player. Player A could not possibly guarantee himself more, since B can keep him down to this value; on the other hand, B cannot guarantee a lower payoff to A since A can get at least the value.

In order to generalize these results to the class of all 2×2 zero-sum games, we use Matrix 3.7a, where the letters may stand for any numbers at all. Assume that we have, as before, put the highest payoff in the upper left, by interchanging rows or columns as neces-

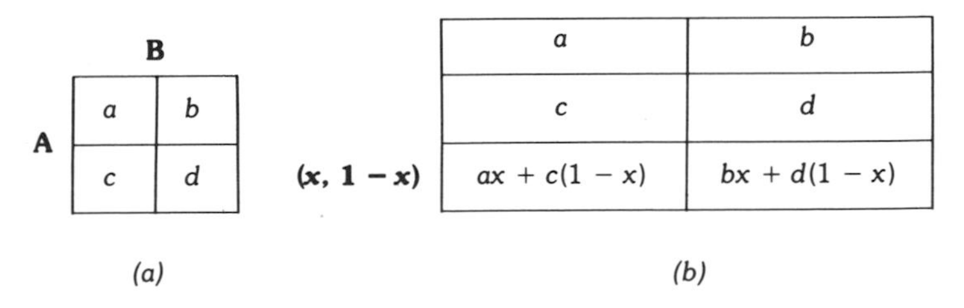

Matrix 3.7 Matrix (*a*) is a generalized 2×2 matrix. Matrix (*b*) is an enlargement of Matrix (*a*) showing a generalized mixed strategy.

sary. Thus a is the largest of a, b, c, and d. Further, assume that d is second largest, since otherwise, as in Matrix 3.5c, there would be a dominant strategy for at least one of the players and hence no difficulty in defining a solution.

The next step is to add a new row to Matrix 3.7a to form Matrix 3.7b. The new bottom row represents the choice to use the mixed strategy $(x, 1 - x)$, which means using the top row with probability x and the second row the rest of the time. The value of x may be anything from zero to one, and our objective is to find just what value of x to use. The expression in the new bottom-left cell is the expected utility when player A uses this mixed strategy and player B picks the left column. The expression in the lower-right cell is found similarly.

We saw in the numerical example that player A could get his best possible guarantee, or maximin, by using a mixed strategy that made B's choice irrelevant, i.e., A got 1 1/4 no matter what B chose. In the general case under consideration here, player A can do the same thing. That is, player A can choose a value of x that forces B to accept a fixed amount of expected utility no matter what he (B) does. To put B in this somewhat "helpless" situation, A equates the two expressions in the bottom row of the matrix:

$$ax + c(1 - x) = bx + d(1 - x)$$

Solving this equation for x yields

$$(a - c - b + d)x = d - c$$

$$x = \frac{d - c}{(a - b) + (d - c)}$$

The maximin mixed strategy for player A is thus to use the top row with probability x, as given by this formula, and the bottom row the rest of the time, that is, with probability $1 - x$.

Notice that in this expression for x, since a and d are the two largest payoffs, it follows that $a - b$ and $d - c$ are positive, so that x itself is positive and less than one, as a probability must be. The formula must not be used when the two largest payoffs are not on a diagonal, for then it will yield a number that cannot be a probability (not in the range from 0 to 1). However, in that event, we have already seen that one player will have a pure strategy that is dominant. This will also be his maximin (pure) strategy.

If player A does use his maximin mixture, his expected payoff will be the expression formed by substituting for x in the expression in either of the lower cells in Matrix 3.7b. This result, which A can guarantee himself, will again be called the value of the game:

$$\text{value} = a\left[\frac{d-c}{(a-b)+(d-c)}\right] + c\left[1 - \frac{d-c}{(a-b)+(d-c)}\right]$$

After some algebraic manipulation this turns out to be

$$\text{value} = \frac{ad - bc}{(a-b)+(d-c)}$$

It is of interest to apply these formulas to Matrix 3.5c. The results are

$$x = \frac{3-(-4)}{(5-0)+[3-(-4)]} = \frac{7}{12}$$
$$\text{value} = \frac{(5)(3)-(0)(-4)}{(5-0)+[3-(-4)]} = \frac{15}{12} = 1\frac{1}{4}$$

These are the same numbers we had before, as they should be. It is even possible to apply the formulas directly to Matrix 3.5a. For example,

$$x = \frac{5-0}{3-(-4)+5-0} = \frac{5}{12}$$

The row containing 0 and 5 is again to be used 7/12 (= 1 − 5/12) of the time. This justifies saying that Matrixes 3.5a and 3.5c are strategically equivalent.

We will show that $(x, 1-x)$, with x as given by the above formula, is the best that player A can guarantee himself. We can then say that this is *what A should do,* and we also know exactly *what it will get him,* namely an expected utility given by the value, $(ad - bc)/(a - b + d - c)$. These two results are the two parts of the definition of *solution* given at the start of Section 3.1. Thus we have found a solution to 2 × 2 zero-sum games with no dominant strategy (of course, for games that do have a dominant strategy, the solution is straightforward, as shown with Matrices 3.1 and 3.2).

We show that $(x, 1-x)$ is indeed the maximin mixed strategy by showing that any other mixture *could* lead to a lower expected utility. First, recall that in Matrix 3.7a we need only consider the case where a is the largest payoff and d the second largest. Now suppose that the probability assigned to the top row were not x but something larger, say $x + e$, where e is a positive number (but where $x + e$ is not larger than 1). Then if B were to use the right column, the payoff to A would be

$$b(x+e) + d(1-x-e) = bx + d(1-x) + (b-d)e$$

which differs from the earlier result in the lower-right corner of Matrix 3.7b only by the extra quantity $(b - d)e$, which is negative because d is the second-largest payoff while b is third or fourth largest. Similarly, by using $x - e$ as the probability of the top row (e again a positive number), we can show that B could use the left column to give A a payoff less than $ax + c(1 - x)$. Therefore, the expression for x obtained above is indeed the optimal probability for A to assign to his top row if the selection principle of maximin mixture is to be adopted.

Continuing the pursuit of a consistent definition of rationality, let us ask whether the following principles are compatible with each other (we repeat 2 here).

◀ **Rationality Principle 2**
Assume that the other player is "rational" and respond optimally to what he will do.

◀ **Rationality Principle 4**
Always use a maximin mixed strategy.

Suppose that you are player B in the general 2×2 game of Matrix 3.7a. In accord with Rationality Principle 2, you assume A to be rational. Then, from Rationality Principle 4, you conclude that he will use his maximin mixture. But against such a strategy it doesn't matter what you (B) do. Therefore, this line of reasoning gives no guidance at all, and certainly cannot be incompatible with anything. Therefore player B can, without being inconsistent, simply follow Rationality Principle 4, using his maximin mixture. (A formula for player B's maximin appears in Section 3.5.3.)

To recapitulate, maximin has several virtues as a decision principle in 2×2 zero-sum games. First, it always makes a specific choice, unlike the notion of dominance, which provides guidance only when there is a dominant strategy. Next, maximin is sensible in that it picks a dominant strategy when one is available and responds optimally in case only the other player has a dominant choice; in all cases, it provides the best security. Finally, unlike maxi*max*, it is compatible with the assumption that all players are guided by the same principles of rationality.

Now that we have a solution, a few last words are in order about some assumptions that underlie it and the interpretation to be made of it. Therefore, the rest of the section touches again on expected payoffs, repeated play, and real players.

We have been treating expected values of payoffs in the same way as ordinary payoffs, and one might well ask whether it is appropriate to do so. After all, the -4 in Matrix 3.5 can be avoided by a pure strategy and not by a mixed one. Suppose a player simply cannot "afford" a -4 payoff? This question raises a false concern and embodies a misunderstanding of the *interval* scale of utilities. The very definition of such a scale is precisely that players' utilities are consistent with the rules of probabilistic expectation (see Section 3.2.2). If a certain outcome of a situation "cannot be afforded," then perhaps its utility should be represented by a lower (or further negative) number.

Now suppose that a particular game is to be played only once. Recall that this problem was set aside earlier in favor of the case of a repeated game. If a game is played only once, it makes little sense to speak of one's opponent deducing a pattern of play. Moreover, if the game is played once, there is little difference between flipping a coin that comes up heads and just deciding to choose heads (in the game of "Matching Pennies"). However, even though a person may play one particular game only once, that person may well play many different games at different times; and if he takes, say, "maximaxing" as a consistent guiding principle, then he can be systematically taken advantage of by an intelligent adversary. Moreover, if we were to postulate maximaxing as a human characteristic, then we would be saying that anyone *can* be systematically taken advantage of (by a "maximax-beating strategy"), but that this is not what happens, since everyone is too busy playing maximax to take advantage of anyone.

In contrast to maximax and the other principles discussed earlier, the principle of maximin mixture can be used regularly (and even publicly announced) without any disadvantage to those who use it. Thus, it is logically consistent to suggest that it might reasonably describe what people would do in playing many different zero-sum games at different times.

All of this does not, of course, mean that all people or even *any* people (other than game theorists) will randomize their choices in the exact proportions prescribed by the maximin mixed strategy. Real people may have different rationalities, and may, moreover, behave "irrationally." Nevertheless, there are two virtues to our argument. First, it shows that it is possible, at least for the type of game considered, to construct reasonable and consistent principles of rationality. Second, it provides a base-line or standard against which to compare what people actually do in such situations. If they deviate from our proposed optimal way of playing, we may ask in what direction and under what circumstances. Such questions are outside the realm of game theory itself, which is a branch of mathematics, but they are certainly appropriate in a study of games as models of social phenomena.

3.5 Solutions to Bigger Games

3.5.1 Winning Strategy Translated from Tree to Matrix

In Section 2.2 we saw that a game of perfect information always has a solution. In the case of games with only wins and losses as outcomes, the solution showed how one of the players could guarantee a win. We also saw that it is possible to translate the tree of a game of perfect information into a matrix. The question naturally arises whether and how the winning strategy that was found for one player in the tree game gets translated into the matrix version of the game. The answer is that it does get translated and it shows up as a strategy that not only is dominant but also contains only wins for the player. Thus, with our usual conventions for zero-sum games, we expect that either some row will have all +1's or that some column will have all −1's, depending on which player has the guaranteed win.

For example, consider "Pile of Four Stones" as represented in Tree 2.3. The strategies for the players can be found as in Section 2.2. Player A has two alternatives, 1 or 2, on the first move. If he picks 2, there are no further decision points for him, so "2" is a strategy. However, if he picks 1, there is the possibility that he may have a second move, and a complete strategy must allow for this possibility. Thus "1, 1" for player A will mean 1 on the first move and then 1 again, should the opportunity for a second move arise. We define "1, 2" similarly.

Player B, on the other hand, will surely only have one move but he does not know which of his decision points he will be at. Therefore he must specify what he will do in either case. His strategies are "1, 1," "1, 2," "2, 1," and "2, 2," where the number before the comma tells what he plans to do if he finds himself at his upper decision point in the tree (if A picks 1) and the one after the comma refers to his lower decision point. Matrix 3.8 shows the various possible results and does indeed have a winning strategy, specifically a row of +1's.

		B			
		1, 1	**1, 2**	**2, 1**	**2, 2**
	1, 1	−1	−1	+1	+1
A	**1, 2**	+1	+1	+1	+1
	2	+1	−1	+1	−1

Matrix 3.8 Matrix translation of "Pile of Four Stones" (Tree 2.3).

3.5.2 Removing Dominated Strategies

In Matrices 3.1 and 3.2 we saw that when a player has only two alternative strategies, one may dominate the other and the dominated strategy can be removed from consideration; the remaining strategy is called the dominant strategy. This terminology can easily be generalized to the case with three or more alternatives. If one strategy dominates *all* others, it is still a dominant strategy. However, with three or more strategies available, yet another possibility arises. Consider, for example, Matrix 3.9, where no row dominates any

	B		
	0	−1	3
A	3	2	1
	4	1	0

Matrix 3.9 No row dominates the others. The center column dominates the left but not the right column. There is therefore no dominant strategy.

other but where the center column dominates the left column. The center column is not, however, a dominant strategy since it fails to dominate the right column. Therefore, we cannot say unequivocally that B should use the center column. What we can say, though, is that B certainly should *not* use the left column. Thus, although we do not find a dominant strategy, we do succeed in removing a dominated one.

Now let us examine A's situation. In Matrix 3.9 player A does not have a dominant strategy and in fact none of his strategies dominates any other (you should check to see that this is true). Now, however, if we ignore the left column, which we can count on any right-thinking player B to do, we find that in the remaining matrix one of the rows is dominated, namely the bottom row. Player A may be tempted to pick it since it contains 4, his best payoff. But it would indeed be wishful thinking to suppose that getting 4 is a serious possibility since getting it presumes that B would pick a strategy that is worse under all circumstances than some other strategy. Thus the bottom row offers only false hopes of such a benefit and can properly be eliminated from consideration.

If it is agreed that the left column and the bottom row are unacceptable, then Matrix 3.10, which includes the remaining two strategies for each player, is all that needs to be considered. In this matrix the two largest payoffs to A are 3 and 2, which are diagonally placed with respect to each other. Therefore no single pure strategy will be optimal for either player, and it is appropriate to compute the

	B	
A	−1	3
	2	1

Matrix 3.10 Refinement of Matrix 3.9, eliminating unacceptable strategies.

maximin mixed strategies as in Section 3.4, using the formula arrived at there. For player A the relevant calculations yield the strategy (1/5, 4/5, 0); the zero indicates that the bottom row is never to be used. If A does indeed mix his choices in this way, he will win an average of 1 2/5 if B uses the center or right column (or any mixture of them) and 2 2/5 whenever B is foolish enough to use the left column. The value of this game, i.e., the amount A can guarantee himself (see Section 3.4), is thus 1 2/5. Player B's maximin mixed strategy is (0, 2/5, 3/5), which guarantees −1 2/5 to B, or −2/5 if A were to use the bottom row. Thus, player B's maximin mixed strategy insures that the payoff to A will be no greater than 1 2/5. Just as in the example in Section 3.4 there is a single number, called the value of the game (in this case 1 2/5), that can be guaranteed by either player. Player A cannot guarantee more to himself, while B cannot be sure of keeping A to less.

The removal of dominated strategies can sometimes be pursued even further. Thus in Matrix 3.11a there is a dominated row

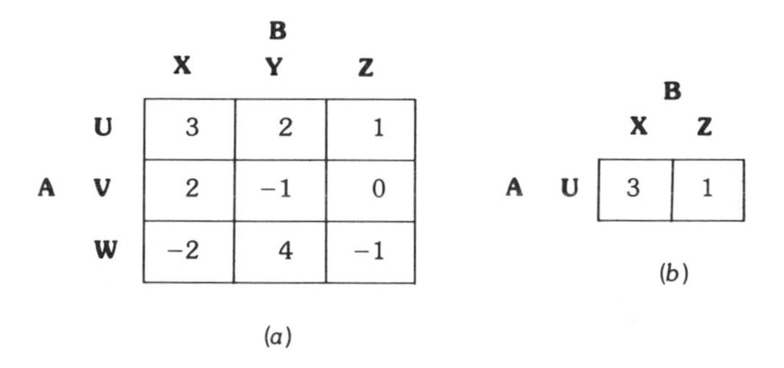

		B: X	B: Y	B: Z
A	U	3	2	1
A	V	2	−1	0
A	W	−2	4	−1

(*a*)

		B: X	B: Z
A	U	3	1

(*b*)

Matrix 3.11 Successive removal of dominated strategies starting with Matrix (*a*) results in Matrix (*b*).

(find it), removal of which leaves a matrix with a dominated column (find it and check to see that it could not have been removed first!). Now another row is dominated, though originally it was not. If you have been stepping right along with the directions, you will now be left with only the top row and the left and right columns. The residual 1 × 2 matrix is shown in Matrix 3.11b. Here only B still has a significant decision, and it is clear that he must pick the right column to get −1, rather than −3. Thus, we anticipate that the outcome of this game will be the upper-right cell with payoff 1 to A and −1 to B.

If all this seems far-fetched, note that this outcome can be supported by another rationale, namely that it is an equilibrium according to the definition given earlier. That is, neither player can do better by single-handedly moving away from this cell. It is not hard to show, though we will not do so here, that whenever removal of dominated strategies "goes all the way," resulting in a single cell, that cell must be an equilibrium.

3.5.3 The Maximin Theorem

Games can be larger than 2×2 and they need not always be reducible by removal of dominated strategies. In the "Old Shell Game" (Matrix 2.2) each player has three strategies and you can easily check that none of them is dominated for either player. The same goes for Game 2.4, "Scissors, Paper, Rock."

Suppose that, what with inflation and all, the old shell game is now to be played for $2 Kewpie dolls with a $1 entry fee. The revised payoffs are shown in Matrix 3.12. No strategy in this matrix

	Showman		
	1	−1	−1
Carnival-goer	−1	1	−1
	−1	−1	1

Matrix 3.12 "Old Shell Game" with revised payoffs.

seems any better than the others, so let us assume that Carnival-goer mixes her strategies equally, using the mixture (1/3, 1/3, 1/3). In that case she will obtain $-1/3$ no matter what Showman does. For example, if Showman uses the center column, Carnival-goer gets $1/3(-1) + 1/3(1) + 1/3(-1) = -1/3$. However, if Carnival-goer uses any other strategy, then Showman can make carnival-goer worse off. Suppose Carnival-goer uses (1/2, 1/4, 1/4), over-using the top row. In response, Showman can simply avoid the left column, say by using the right column. Then the payoff to Carnival-goer will be $1/2(-1) + 1/4(-1) + 1/4(1) = -1/2$, which is worse than the $-1/3$ that Carnival-goer can insure by the more prudent equal-split mixture.

The (1/3, 1/3, 1/3) mixture in this game is a maximin mixture and the resulting payoff to Carnival-goer of $-1/3$ is the value of the game. (We just happened upon the right mixture here, but in general we have not solved 3×3 games. Nevertheless, perhaps you can guess how Showman can guarantee that the payoff to Carnival-goer will be $-1/3$. That is, what is Showman's maximin mixture?) We have now seen a variety of games each with a value that can be guaranteed by either player. Each is an example of a more general result, the Maximin Theorem. This is the central theorem of two-person, zero-sum games with any finite number of strategies for either player. We shall state it without proof:

◀ There is a number, called the value, and a mixed strategy for player A that guarantees him at least the value. Also, there is a mixed strategy for player B that guarantees that A gets at most the value.

We came close to proving this for the 2 × 2 case by using Matrix 3.7 in Section 3.4. Pursuing the same line of reasoning for player B that was used there for player A, we find that if $(y, 1 - y)$ is used by B as a mixed strategy, the resulting payoff (to A) will be

$$ay + b(1 - y)$$

if A uses the top row. If A uses the bottom row, the payoff will be

$$cy + d(1 - y)$$

Equating these two expressions and solving for y gives

$$y = \frac{d - b}{(a - c) + (d - b)}$$

Substituting this expression for y into the first of the two payoff expressions above (either will do, since they were set equal), yields a payoff of

$$a\left[\frac{d - b}{a - c + d - b}\right] + b\left[1 - \frac{d - b}{a - c + d - b}\right] = \frac{ad - bc}{a - c + d - b}$$

This same expression was found in Section 3.4 and was called the value because player A could assure himself of getting *at least* that much. Similarly here, we find that B can insure that A gets *at most* that much. Thus the proof of the maximin theorem is complete in the 2 × 2 case. Proof of the theorem for larger games (more alternatives) may be found in Luce and Raiffa (1957). In lieu of a proof, the following subsection provides some intuition about some larger games.

3.5.4 Many × Two Games

If a player has three (or more) strategies, one of them may be dominated by a mixture of two (or more) others. For example, in Matrix 3.13 strategy V is dominated by (1/2, 0, 1/2), the mixed strategy of

		B X	B Y
A	U	10	2
A	V	3	3
A	W	2	10

Matrix 3.13 In this matrix one strategy (V) is dominated by the mixed strategy of choosing one of the two others (U, W), each with probability 1/2. Therefore, V is deemed unacceptable even though it is not dominated by any *pure* strategy.

picking U or W each with probability 1/2. This is true even though no pure strategy dominates any other.

Additional insight into this situation is obtained by representing each of player A's strategies as a point on a graph. Thus, for strategy U player A will get a payoff of 10 or 2, depending on whether B uses X or Y. We therefore represent U as the point 10,2, (an x-coordinate of 10 and a y-coordinate of 2). The points V and W can be similarly plotted. All this is done in Graph 3.1.

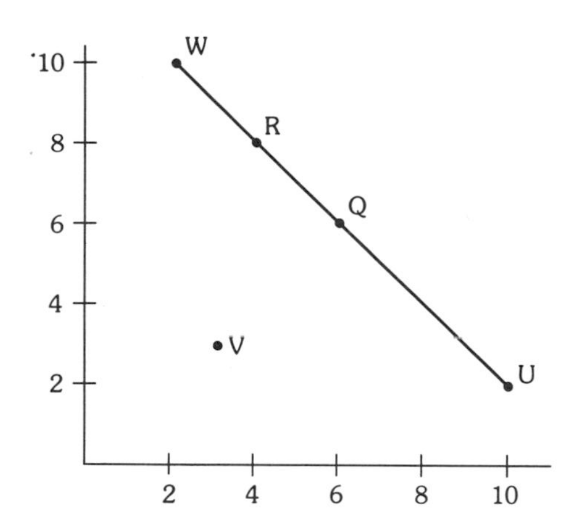

Graph 3.1 A coordinate representation of the mixed strategies available to player A in Matrix 3.13. In such a representation any strategy that is "northeast" of another (as Q is to V here) dominates it.

The strategy (1/2, 0, 1/2) yields an expected payoff of 6 whether X or Y is used, as you should verify. This strategy is represented as the point Q on the graph (coordinates 6,6). Q is to the upper-right of V, or by analogy to a map, we might also say that Q is "northeast" of V. In general, whenever one point is northeast of another, the northeasterly strategy dominates the southwesterly strategy. Domination is thus translated into "northeasterliness."

There are other possible mixtures of strategies W and U, and each of them corresponds to a point on the line connecting points W and U. For example, the strategy (1/4, 0, 3/4) has payoffs 4 and 8 when B uses X and Y, respectively. The appropriate point (4,8) is labeled R on the graph and does indeed lie on the line from U to W. Also, R lies closer to W than to U, which reflects the fact that it is weighted more heavily (3/4) with W than with U.

To consider all mixtures of U and W, we examine $(a, 0, 1 - a)$. Then if player B uses X, the payoff will be

$$X = 10a + 2(1 - a) = 2 + 8a$$

while if B uses Y, the payoff will be

$$Y = 2a + 10(1 - a) = 10 - 8a$$

Eliminating a from these equations (solving one for a and substituting the result in the other) yields

$$Y = 12 - X$$

which will be recognized as the formula for the straight line with intercept 12 and slope -1. Notice, however, that while the straight line given by $Y = 12 - X$ extends indefinitely in either direction, our mixed-strategy points cover only the segment between U and W. (Points outside this range correspond to values of a outside the interval 0-1, and these are not appropriate since a is a probability.)

Now consider the 5×2 game of Matrix 3.14a. The payoff

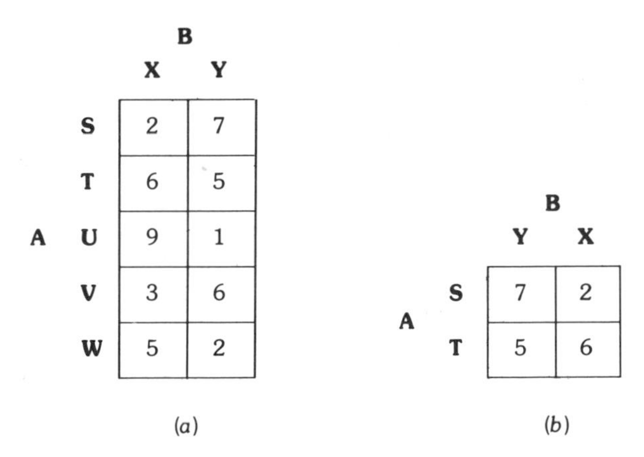

(a)

A	B: X	B: Y
S	2	7
T	6	5
U	9	1
V	3	6
W	5	2

(b)

A	B: Y	B: X
S	7	2
T	5	6

Matrix 3.14 Matrix (*a*) is a 5×2 game. Matrix (*b*) shows only the two pure strategies for player A that are included in his maximin strategy. (Note that, by convention, the columns have been interchanged from Matrix *a* to show the highest payoff in the upper-left cell.)

pairs for each strategy are plotted as points in Graph 3.2, along with the (dotted) line X = Y, which passes through the points (0, 0), (1, 1), (2, 2), (3, 3), etc. To see the significance of such a line, recall that in 2 × 2 matrices the maximin mixture for player A always had the same payoff for the left column as for the right column. Thus its point would have to lie on this dotted line.

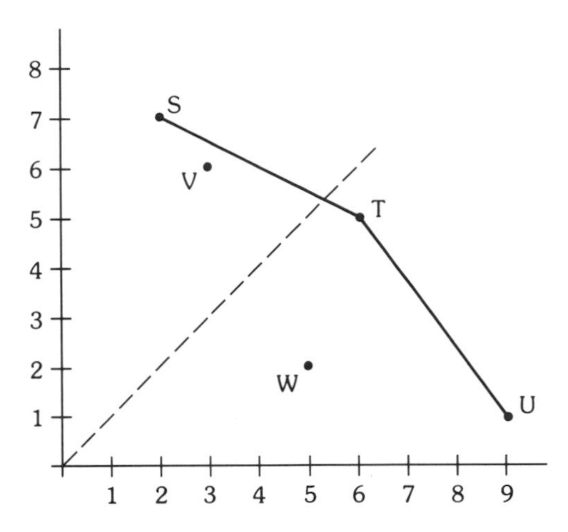

Graph 3.2 Coordinate representation of the strategies available to player A in Matrix 3.14*a*.

In Graph 3.2 it leaps to the eye that W is southwest of T and hence is dominated by it. The strategy V is also dominated, not by any pure strategy but by some point on the line connecting S and T, that is, by some mixture of S and T. Removing dominated strategies leaves only S, T, and U. (We do not show the line connecting S and U since all these points are southwest of points already shown on the two segments between S and T and between T and U.)

Considering now only points on the "northeast frontier" running from S to T to U, we see that one such point has X = Y. This is the point where the dotted line crosses the line segment S–T, with coordinates (5 1/3, 5 1/3). We thereby deduce that a judicious mixture of strategies S and T by player A will guarantee an expected payoff of 5 1/3. We will see that this is the value of the game. The fact that the point (5 1/3, 5 1/3) is closer to T than to S is a graphical reflection of the fact that the maximin mixed strategy uses a higher probability of T than of S.

To confirm these claims, we now find the maximin strategy in the game matrix by ignoring the bottom three rows and using the maximin formula for 2 × 2 games on the remaining part of the matrix. This remaining part is shown as Matrix 3.14b, with columns interchanged in order to meet our convention of having the highest payoff in the upper-left cell. The maximin mixed strategy is

$$x = \frac{d - c}{a - b + d - c} = \frac{6 - 5}{7 - 2 + 6 - 5} = \frac{1}{6}$$

and $1 - x = 1 - 1/6 = 5/6$. In the original 5×2 matrix this mixture is (1/6, 5/6, 0, 0, 0) with U, V, and W each used with probability 0, that is, never. The resulting payoff if B uses X is $1/6(2) + 5/6(6) = 5\ 1/3$.

Thus, 5 1/3 is the value of the game. It can be guaranteed by A by using the mixture of S and T computed above, and there is no way for A to guarantee more. This is verified by noting that for player B the strategy mixture (1/3, 2/3) insures that A will receive no more than 5 1/3.

A 3×Many game can be handled similarly to the 2×Many, but it is awkward to draw the needed three-dimensional graph. Still more awkward are the four-, five-, and higher-dimensional graphs needed for even bigger games. What can be done is to translate our graphical technique to equivalent algebraic manipulations, which can be more readily extended as matrix size increases. (We shall not pursue this topic any further, but it turns out that there is an elegant way of solving zero-sum games that also solves the so-called linear programming problem, which finds application in business management. See, for example, Singleton and Tyndall, 1974.)

Exercises

○ **1** **(a)** Two unbiased coins are flipped. What is the probability of getting

(i) Both heads?
(ii) Both tails?
(iii) One of each?

(b) What is the expected value of the payoff, in cents, for the following situation. A die is rolled and if the number 6 turns up you get 25¢; if an odd number shows up you get 5¢; otherwise you lose 10¢.

○ **2** What information do you need in order to say that Timothy's utilities for A, B, and C are 1, 2, and 5, respectively?

○ **3** Merv is indifferent between outcome B and a lottery in which outcomes A and C are possible, A having probability 1/3. He prefers A to C.

(a) Can we represent these facts by saying that for Merv, A has utility 4, B has −2, and C has −5? (Yes or no?)

Suppose we say that for Merv, A has utility 9, B has 5, and C has 3.

(b) If Merv is indifferent between D and a lottery in which A and C occur with equal probability, what is the utility to him of D?

(c) If Merv is indifferent between A and the lottery (1/2 C, 1/2 E), what is the utility of E?

(d) Suppose in addition to all of the above, Merv also says that he prefers the lottery (1/3 E, 2/3 C) to the lottery (1/4 A, 3/4 B). Is this consistent with Merv having an interval scale of utilities?

○ **4** In each of the following three cases an individual prefers A to B and B to C. Moreover, each person is capable of lottery judgments; that is, each has an interval scale of preference. Restate each of the following statements as a preference or indifference between B and a lottery involving A and C. For each, state the probabilities involved in the lottery.

(a) Wendy's preference for A over B is four times her preference for B over C.

(b) The average of Xenophon's utility for A and for C exceeds his utility for B.

(c) Yuri is indifferent between B and a 50-50 lottery between A and J. He is also indifferent between J and a 50-50 lottery between B and C.

○ **5** For the zero-sum game shown below:

(a) Find the largest number in each column and see whether it is also the smallest in its row. That is, find all equilibrium points.

(b) Find the smallest number in each row and write it at the right side of its row. These are the "row minima."

(c) The largest of these row minima is Row-player's maximin (maximum of the minima). Row-player can guarantee that he will get at least this amount. The row containing it is his maximin alternative. Find these.

(d) Write each "column maximum" at the foot of its column.

(e) The lowest of these is Column-player's minimax (minimum of the maxima). Column-player can guarantee that he will do no worse than this value with its sign reversed (positive to negative, or vice versa). The column containing it is his minimax alternative. Find these.

(f) Suppose players use maximin and minimax alternatives, respectively, in the accompanying matrix. What is the result?

1	−2	3
0	4	−4
−3	1	6

○ **6** For each of the following games, (i) use the appropriate formula to compute the maximin mixed strategy for A, and (ii) use the result of (i) to find the expected payoff to A if he uses that strategy while B uses the left column. Check your answer using the formula for the value of the game.

(a)

A \ B		
	1	−1
	−1	1

(b)

A \ B		
	2	−1
	−1	1

(c)

A \ B		
	10	−1
	−1	1

(d) In ordinary, everyday, intuitive terms, why does player A go after the 10 in game (c) so rarely?

○ **7** Consider Matrix 3.2:

(a) When there is a dominant strategy, the formulas for maximin mixture become meaningless. Show what happens when you try to use those formulas for each player in Matrix 3.2.

(b) Find the maximin pure strategy for each player.

(c) Show that player B's maximin is also dominant.

(d) Show that A's maximin is the best response if it is anticipated that B will use his dominant strategy.

○ **8** A puts down two pennies while B puts down one; neither shows the other what he is doing. When the pennies are uncovered, A wins 1¢ if there are a total of two heads and one tail; B wins 1¢ if there are two tails and a head; no money is exchanged if all three coins face the same way.

(a) Give the matrix.

(b) One of the players has an alternative that is dominated. Draw a line through it.

(c) Solve the remaining game.

○ **9** **(a)** In Matrix 2.2, if Carnival-goer restricts herself to mixtures of only two alternatives, what is the most she can guarantee herself?

(b) Show that she can guarantee more—than in part (a)—by some mixture of all three rows.

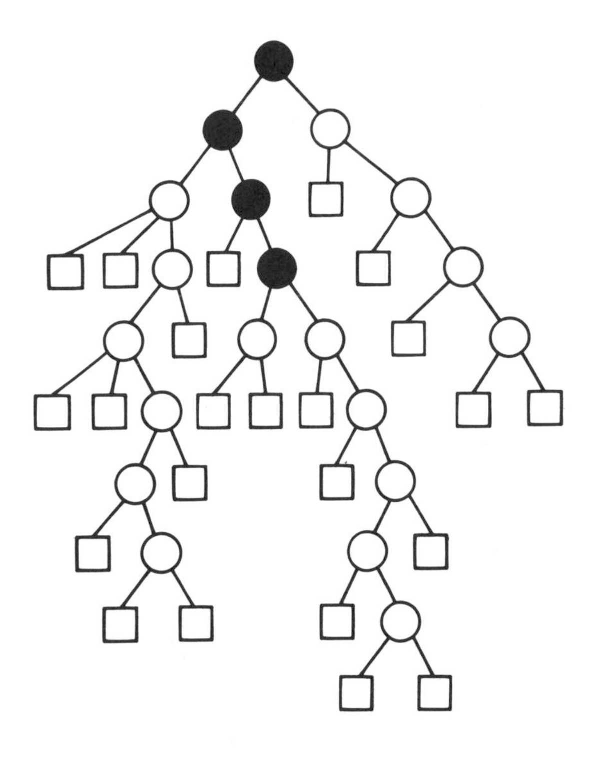

4 Mixed-Motive Games for Two

In this chapter we enter the realm of variable-sum games, but the concepts introduced in connection with zero-sum games will remain useful. In particular, we shall continue to refer frequently to trees, matrices, dominance, equilibrium, mixed strategy, solution, and rationality. Some of these words will retain their original meanings but others will not. Mixed strategy, for example, has the same meaning as before and will be important for our study of enforcement in Section 4.1. In general, however, mixed strategies will not play the central role they did in Chapter 3, where they were crucial to the maximin solution. The word "solution" continues to be used but in quite a different sense than previously. Solutions, to be examined in Chapter 6, will involve not only a competitive but also a cooperative aspect, hence the term "mixed-motive games."

"Equilibrium" retains its meaning but takes on greater significance in the discussion of "deficient" equilibria in Section 4.2. Such outcomes are undesirable to both players, something not possible in zero-sum games. Although both players in mixed-motive games can do better by simultaneously moving away from such outcomes, they can do so only by a coordinated effort in which each puts himself at the mercy of the other. Such is the paradox of "Prisoner's Dilemma," a game in which a group version of "rationality" is at odds with the individual-oriented rationality principles of the preceding chapter.

Another new aspect of equilibrium points is that some are vulnerable to threats, and this vulnerability is the subject of Section 4.3. The credibility of a threat will dramatically influence the solutions proposed in Chapter 6 for mixed-motive games.

The large number of very different games in this chapter will, it is hoped, suggest the rich variety of decision-making considerations involved in human interaction. This variety will appear all the richer when, in the next chapter, the trees and matrices are subjected to the circumstances governing their use.

4.1 Complying, Violating, and Enforcing

> To control automobile speeding within a 65-mile-an-hour limit or to enforce a $1.15 minimum wage provision are concrete administrative objectives. In these and similar cases the administrator's problem is to maximize compliance and cooperation, for there will always be resistance, footdragging, and some overt defiance. If there were no resistance, there would be no need for the administrative program at all; and if it were impossible to increase the frequency of compliance through governmental action, there would equally obviously be no reason for administrative action. [Edelman, 1964, p. 44]

Regulations and their enforcement are an issue at all levels of social organization. Nations have laws governing both individuals and corporations; states have rules for cities and both have laws governing their citizens; businesses have rules for employees, schools have rules for students, and organizations have rules for their members. At the international level the momentous Strategic Arms Limitation Talks (SALT) have been vitally concerned with, and frequently snagged on, questions of compliance and enforcement.

4.1.1 Speed Limit

Let us consider the case of traffic-law violation, or the game "Speed Limit." We ask first who the players are. There are two (in one possible view of the situation): Driver, deciding whether to violate or comply with the law; and Enforcer, deciding whether to enforce or not. But who is Enforcer? Shall we say the police officer? The entire police force? The city (or state)? The officer cannot really be regarded as a decision-maker and therefore, by definition, a player; he is supposed to enforce whatever violations he sees, or whichever ones exceed some degree of flagrancy. Therefore, let us assume that the city is Enforcer. It is the city that must decide how many traffic police man-hours per year both are needed and can be afforded. This decision will determine two things: a cost in salary dollars and a certain probability that any given intersection or stretch of road will be under surveillance at any point in time. This probability reflects Enforcer's mixed strategy, a compromise between no enforcement and the (fiscally) unrealistic strategy of constant enforcement everywhere.

What are the payoffs in this situation? To answer this question, we ask what alternatives each player has and why he would pick one alternative rather than another. Why does someone speed? Why does someone else stay under the speed limit? Why does the city put police on the streets? Why not more of them? The answers suggest the payoffs: time, money, injury. Driver wants to save time but not pay a fine and certainly not get into an accident. (A driver who believes that one should always obey all laws is not affected by enforcement and thus apparently has no decision to make. He is therefore of little interest to us here. Likewise, a driver who enjoys speeding for its own sake, i.e., at any cost, also has no decision to make and therefore does not concern us.) Enforcer wishes the other drivers and pedestrians not to be injured (and perhaps also wishes them not to have to endure the fear of injury) but must not spend excessively in the effort to avoid injuries.

These various aspects of the payoffs are collected together in Matrix 4.1. The lower-right corner is taken as a reference point and assigned a payoff of 0 for each player. This simply means that other outcomes are evaluated in comparison with this one. We assume that payoffs are interval-scale utilities, as in Section 3.2.2. For example, if Enforcer is following the strategy "Ignore" (not enforcing), then Driver may shift from compliance to violation with a

		Enforcer	
		Enforce	Ignore
Driver	Violate	$t - s - f, -h - b$	$t - s, -h$
	Comply	$0, -b$	$0, 0$

Matrix 4.1 "Speed Limit" in general terms.

resulting increase of $t - s$, i.e., the time he saves is worth t, while the loss of safety is worth $-s$. The corresponding hazard to others, $-h$, is a cost to Enforcer and hence is put after the comma in the matrix cell. The term b represents the cost to Enforcer of the policeman's time and equipment and has to be paid when enforcement is chosen, whether this choice has any effect on Driver or not. Finally, the fine, $-f$, occurs in the one cell where violation meets enforcement. The upper-left cell involves some combined effects.

To get a clearer sense of this matrix, it will be helpful to put in numerical payoffs. Some costs and benefits are easier to evaluate than others. For some we can use money as a common measure to aid our thinking. It is easy to find out the budgeted costs of enforcement. We might also agree to convert the driver's time-saving into money by using his earning rate per minute at his job. An injury can in part be translated into time lost from work plus medical expenses, but pain and inconvenience are much harder, perhaps impossible, to put a price tag on. In fact, evaluations of such losses by the courts vary tremendously.

Having mentioned these difficulties, we now set them aside and pick some numbers (Matrix 4.2). These numbers are not en-

		Enforcer	
		Enforce	Ignore
Driver	Violate	−190, −25	10, −5
	Comply	0, −20	0, 0

Matrix 4.2 "Speed Limit" with specific numerical payoffs.

tirely arbitrary since they do conform to Matrix 4.1. Thus the number after the comma in the upper-right cell must be negative, since it is $-h$. Further, the fact that payoffs −5 and −20 add up to −25 is consistent with $-h$, $-b$, and $-h - b$ in Matrix 4.1. The large value 200, for f, in comparison with $t - s = 10$ (so $t - s - f = -190$), reflects the fact that, for Driver, getting caught once outweighs several undiscovered violations.

In Matrix 4.2 Driver does not have a dominant strategy but Enforcer does (both parts of this statement can even be made about Matrix 4.1, provided that $t - s$ is positive but is outweighed by f). Thus, the simplest analysis suggests that Enforcer will ignore a violation, which in turn makes it advantageous for Driver to violate. These choices are in accord with Rationality Principles 1 and 2 in Section 3.3. The resulting payoffs would be 10 for Driver and −5 for Enforcer, but in fact Enforcer can do better than this.

If Enforcer chooses a mixed strategy of 10 percent enforcement, announces it, and sticks to it, then Driver faces a payoff matrix consisting of a 10–90 percent mixture of the two columns of Matrix 4.2. This situation is shown in Matrix 4.3. (The computation of

		Enforcer (0.1, 0.9)
Driver	Violate	−10, −7
	Comply	0, −2

Matrix 4.3 Driver's options in "Speed Limit" if Enforcer uses a mixed strategy with probability of enforcement equal to 0.1.

expected utility used to obtain the numbers in Matrix 4.3 is similar to that for Matrix 3.6. For example, the −7 results from $(0.1)(-25) + (0.9)(-5) = -2.5 - 4.5 = -7$.) Faced with these alternatives, Driver will do better to comply, taking 0 rather than −10. The associated payoff for Enforcer is −2, a better result than he could hope for with a consistent (pure) strategy of either enforcing or ignoring.

In this mixed-strategy analysis we have allowed special consideration to Enforcer. He gets to preempt the situation by going first and announcing his choice. Why not allow this privilege to Driver? Suppose that a driver writes to City Hall announcing his intention to drive 10 miles per hour over the speed limit, cops or no, and stating that there is no point in enforcing. But even though our matrices here have only two players, Enforcer and Driver, the fact is that Enforcer is simultaneously playing this game with a multitude of drivers, and is hardly affected in reality by any single one of them. His policy must be determined by the overall results, that is, the sum of the payoffs from this multitude of games. He is therefore unresponsive to any proclamation made by a single driver.

Once we admit that in reality there are many Driver players in the game "Speed Limit," Matrix 4.2 is called into question; it has a single set of payoffs whereas different drivers may have different utilities—some may be more desperate to save a few minutes, or less upset at the prospect of being fined than others. From Matrix 4.1 it follows that a mixed strategy with probability of enforcement of just a little more than $(t - s)/f$ will deter a driver.* Enforcer can figure the results of various strategies without knowing the details of every driver's utilities, provided he has a general idea of how large the value of $(t - s)/f$ tends to be (more precisely and technically, how it is distributed over the population). In practice, such information is found by setting the enforcement budget at a certain level and seeing whether the rate of compliance is satisfactory. If one imagines starting the budget at zero and gradually increasing it, compliance should increase as more and more individuals' value of $(t - s)/f$ is surpassed. After a while, diminishing returns will set in and an equilibrium point will be established at which additional expenditures on enforcement are not worthwhile, i.e., the increase in the number of individuals whose decisions are changed to compliance becomes proportionately more costly.

4.1.2 The Fourth Amendment and Mixed Strategy

The traffic officer's "spot check" is not the only use of a mixed strategy in enforcement. On January 21, 1974, the U.S. Supreme Court accepted for argument a case "challenging unannounced pollution inspections at industrial sites as unreasonable searches in violation of the Fourth Amendment. The state of Colorado filed the appeal after a state court invalidated the inspection procedures" (*Los Angeles Times,* 1/22/74). What had happened here is that an environmental standards agency of the state of Colorado had used a mixed strategy of enforcement, and for the same reasons as those of the traffic law enforcer. The use of such a strategy involves an element of surprise. While the state is perfectly willing to let the proportions of its strategy mixture be known, it certainly must not let potential violators know in advance that they have (randomly) turned up for inspection at a specific time. At least such information must not be available far enough in advance for a polluter to "clean up his act" just in time for the inspector, only to resume polluting afterwards.

* If Enforcer uses the mixture $(x, 1 - x)$, then Driver has a choice between 0 for compliance or an expected utility of $x(t - s - f) + (1 - x)(t - s)$ for violation. This latter expression will be less than 0 just in case x exceeds $(t - s)/f$.

Why has the Colorado Supreme Court disallowed this environmental inspection system? The element of surprise, which is necessarily associated with a mixed strategy, is not at issue. The courts have had no qualms about allowing evidence gathered by police who park partly hidden at random times at key speed traps or stop signs and wait for violators. What possible basis is there for distinguishing the two cases?

Spot checks of traffic differ from spot checks of industrial plants in at least two ways. First, the streets are public whereas factories are semiprivate (hence, the possibility of "unreasonable *search*"). Second, specific plants are preselected for scrutiny whereas individual drivers are not. The latter distinction is a simple consequence of the fact that factories are stationary so you cannot randomize them by waiting on the street corner for one to drive by. However, if each has an equal likelihood of being inspected, there can be no complaint of discrimination.

Thus the legal issue apparently centers on the entry of a building. But this aspect of the situation is something of a technological accident. One could imagine helicopter-borne detection equipment hovering briefly over a chimney and obtaining the needed information without entering the building. It is even conceivable that long-distance spectrographic chemical analysis could be performed without entering the plant's air-space. Whether or not such technology is currently practicable, there is a clear parallel between violations of public atmosphere and violations on public roads. The two situations have analogous decision structures and both call for a mixed strategy of enforcement.

There is, of course, another approach to discouraging violations of various laws and regulations, namely making the penalties higher. As I am writing, a bill is just entering the California legislature that would increase the penalty for certain kinds of business fraud from $1,000 to $50,000. The announced intention of the bill is to stop repeated offenders from regarding the fine as just one of the costs of doing business. Without judging this particular proposal, we may note that its reasoning carried to the logical extreme suggests the death penalty for running a stop sign. This is "effective deterrence," but is also clearly "cruel and unusual punishment," and hence is ruled out by the Eighth Amendment (not to mention common sense).

Additional topics arise naturally in connection with this sort of analysis. For example, how abhorrent is it that a perpetrator of fraud regards fines as business costs? Surely the speeder follows a similar principle of risk rather than absolutes or moral imperatives. The debate between pollution standards and pollution taxes is put on just these terms when one argues, as many a legislator has, that a pollution tax gives official blessing to any amount of pollution. Such an argument loses force if both tax and fine are regarded as business costs equivalent to any other monetary costs. Once the problem is reduced to one simply of costs, the economist will be happy to explain the potential effectiveness of a tax. Pollution taxes are discussed at some length in Section 6.2.

Recall now our assumption in "Speed Limit" that no single potential violator can affect the decision of the enforcer. This aspect of the situation is reminiscent of the economist's "perfect market," in which all buyers and sellers do business at the current equilibrium or market price; no single one of them affects that price. However, the situation changes radically when a multitude of individual decision-makers is replaced by one or a few; we then have the study of monopoly, monopsony, oligopoly, and large labor unions, or of organized civil disobedience, mutiny, and prison uprisings. If potential violators act in concert, they can indeed affect the decisions of the enforcer. For example, the chief warden of a jail may have a system of regulations, a level of enforcements, and a set of penalties that he declares to be unchangeable. But let the prisoners band together and simultaneously violate those regulations, and there will be a whole new pattern of decision-making.

4.2 Five Symmetric Games

Offhand it may seem unlikely that symmetric 2×2 games, the topic of this section, could hold much interest. After all, such games present only two options to each player, and the same two at that. Nevertheless, we will find that such games lead us to a rich variety of strategic considerations and are applicable to momentous real-world decision structures.

4.2.1 SALT and Prisoner's Dilemma

From the point of view of game theory the Strategic Arms Limitation Talks (SALT) provide a curious situation indeed—the two players are negotiating the rules of the game they will play. The United States and the Soviet Union might, for example, agree upon rules for inspection of potential weapons production sites, so that each nation would have the right to make a certain limited number of inspections per year. Then each nation would be an enforcer, possibly employing a mixed strategy like that used by the traffic officer or the Colorado pollution inspector, and each would at the same time be a potential violator deciding whether to risk being caught.

The catching of a violator presents a serious problem in the diplo-military situation, since neither side has the power to enforce a penalty on the other. Thus a violation may throw them back into the *negotiations game,* discussing rules for a new arms-and-inspection game. Changes in technology also require new negotiations.

The reason why the negotiations are of such great concern, even though progress in them is painfully slow, is that if they break down, the two nations are automatically engaged as players, not rule-makers, in the game of "Arms Race," which typically has two losers and no winners. We shall take a more detailed look at the "balance of terror" in Section 4.3.2, but some important considerations can be made clear with the simple model that follows.

Game 4.1 Arms Race

Each player chooses the size of his annual arms budget. If one player has greater arms strength than the other, he gains a diplo-military advantage. However, resources devoted to arms are taken away from other goods and services.

To reduce this game to a small matrix, we assume that each player has only two alternatives: high or low arms expenditure. This approach yields Matrix 4.4a, which shows the possible results. The assumption that gaining a diplo-military advantage or avoiding a disadvantage is more important to each nation than the equal dollar value of goods and services leads to Matrix 4.4b. The payoffs in this matrix are given on an ordinal scale (Section 3.2.1), reflecting our very minimal assumptions. The terms B, S, T, and W denote the Best, Second, Third, and Worst outcomes, respectively, to each player. The ordinally specified Matrix 4.4b can be taken as the definition of the game "Prisoner's Dilemma."* It can be regarded as a model not only of arms races, but of price wars, legislative log-rolling, and a variety of other situations. The multi-player gen-

* The name "Prisoner's Dilemma" comes from a classic case in which two suspects are taken into custody and separated. The district attorney tells them that he does not have adequate evidence to convict them in a trial and presents them with the possible outcomes. If one confesses and the other does not, the confessor will be released for turning state's evidence (the best outcome, B) while the other will have the book thrown at him (the worst, W). If neither confesses he will have them both booked on a trumped-up minor charge (second best, S). If both confess, they will both be prosecuted but he will recommend leniency (the third best outcome, T). The choice to confess corresponds to the choice of "high" in Matrix 4.4b.

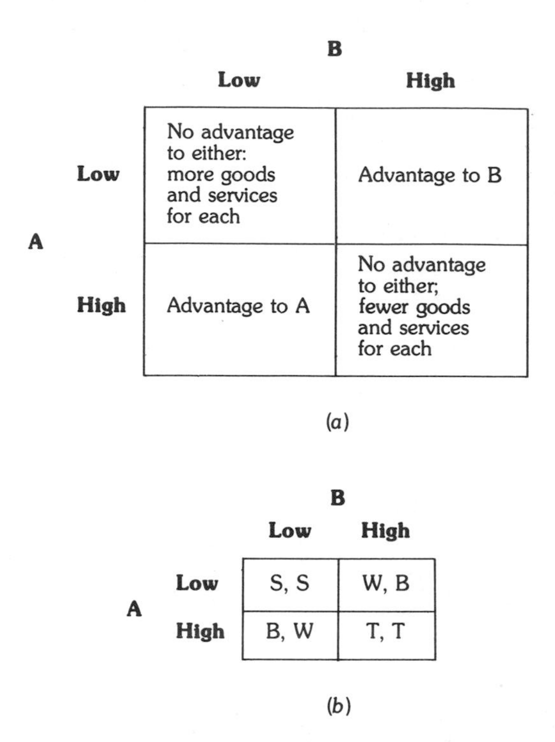

A \ B	Low	High
Low	No advantage to either: more goods and services for each	Advantage to B
High	Advantage to A	No advantage to either; fewer goods and services for each

(*a*)

A \ B	Low	High
Low	S, S	W, B
High	B, W	T, T

(*b*)

Matrix 4.4 "Arms Race." Matrix (*a*) shows the real-world outcomes; Matrix (*b*) indicates players' preferences among those outcomes.

eralization of it has literally dozens of applications, several of which are treated in Chapter 7.

The strategic properties of this matrix present a dilemma that has been the source of seemingly endless strategic and moral debate. Briefly put, the paradox of "Prisoner's Dilemma" is as follows.

☐ **1** Each player has a dominant strategy.
☐ **2** If both players use the dominant strategy, the result is worse for *both* players than if both had instead used their *dominated* strategy.

Specifically, the dominant strategy for player A is the bottom row; in the first column payoff B (Best) is better than S (Second), while in the second column he prefers T (Third) to W (Worst). Player B also has a dominant strategy, preferring each result in the right column to the corresponding one on the left.

Both of these dominant choices, the bottom row and right column, are the "high" arms budget. If both players use this dominant strategy, they end up with (T, T), which means both of them are spending a lot of money on armaments without it doing either of them any good. The irony of this result is that both players would be better off if both did not choose their dominant strategies, in violation of Rationality Principle 1 (Section 3.3). In other words, if both pick "low," the payoffs are (S, S), which both prefer to (T, T) because they are still equal in arms but at a lower cost.

This breakdown in the usefulness of Rationality Principle 1, or dominance, is of the utmost significance. Two points about it deserve particular attention. First, part 2 of the paradox of "Prisoner's Dilemma" cannot possibly occur in zero-sum games, since in such games both players cannot possibly improve their lot simultaneously. Second, and perhaps more important, recall that dominance is not just any old principle. It is stronger than the principles of maximin, maximax, maximum average, or minimum possible regret in the sense that if a strategy is dominant it will meet all those other criteria as well. In contrast, a strategy may be, say, maximin without being dominant (see the discussion of Matrix 3.4).

A good way to look at this dilemma is by trying again to define "rationality." Suppose two players arrive at some outcome and *together* they have the power to switch to some other outcome that is better for both of them. Then we might say that "group rationality" calls for them to switch, that is, not to settle on the outcome that is worse for both of them. If we refer to picking dominant strategies as "individual rationality," then we get the following paradoxical result: In a "Prisoner's Dilemma" game, *individual rationality leads to group irrationality.*

4.2.2 Preference Order, Symmetry, and Separability

Suppose you run a gas station and are about to start a price war. Suddenly you learn that the paint you need for the signs announcing your cut-rate prices has gone up by an inconsequential amount, say 1¢ per quart. Strictly speaking, the payoffs have changed. Nevertheless, the *relationships among the payoffs* are still the same. How can we express the idea that you are still playing essentially the same game with other gas-station owners? A good way is to express payoffs on an ordinal scale of the sort discussed in Section 3.2. Then if there is a small change in any of the payoffs that does not change their order of preference by any player, the game is still the same.

Each of the five games to be discussed in this section has its own ordering of payoffs, different from the others. Since only the preference order is being considered, we shall simply refer to a player's payoffs as B, S, T, and W, for Best, Second, Third, and Worst outcomes, respectively. Even considering only preference order, there is still a vast array of distinguishable games. For the smallest possible matrix size, the 2 × 2 game, there are 78 significantly different ways to distribute the letters B, S, T, and W in the matrix for the two players.* The particular five games discussed in this section have been selected from among these 78 possibilities.

In exploring the strategic properties of these five games, it is important to notice that they are all *symmetric.* A symmetric game is one that looks the same to both players. (Some nonsymmetric games are discussed in Section 4.3.) As an example, consider the "Prisoner's Dilemma" game of Matrix 4.5. If both players pick strategy I, both get 3; if they both pick II, both get 2. That is, if they both do the same, they both get the same. This is the first part of symmetry.

		B: I	B: II
A	I	3, 3	1, 4
A	II	4, 1	2, 2

Matrix 4.5 "Prisoner's Dilemma," an example of a symmetric game.

Now look at what happens if they pick differently from each other. In that case, one will get 4 and the other will get 1. But notice that the 4 always goes to the II-chooser. If A picks II, the 4 in the lower-left cell is before the comma so it is A's payoff; if B picks II, the 4 in the upper-right is after the comma so it goes to B. In sum, a symmetric game is one where the payoff is determined by what you do, not by who you are. If you know that Sean picked strategy I and Dawn picked II, then without knowing whether Sean is player A or player B, you can be sure that Sean's payoff is 1 and that Dawn's is 4.

Like symmetry, the notion of *separability* is an important property of some games. Some games can be separated into two parts and this separability can be helpful in understanding their strategic properties. Take an example:

● **Game 4.2 Noisy Neighbors**

Horace and Boris live next door to each other in an apartment building with thin walls. Horace plays ping pong every Friday and

* This result is from Rapoport and Guyer (1966). They discuss many of the games in strategic terms and introduce, among other things, the definition of a threat-vulnerable equilibrium used in the next section.

this is a minor disturbance to Boris. If Horace plays, the resulting payoffs are 3 to Horace and −1 to Boris, but they are 0 and 0 if he does not. Boris, on the other hand, has political debates every Saturday with comparable resulting payoffs: 3 to himself and −1 to Horace, or, if he refrains, 0 and 0.

One playing of this game will be regarded as a single weekend, that is, one decision by each person. Also, we assume that the payoffs from Friday and Saturday can be added. Thus, if both players engage in their disturbing activities, each gets 3 − 1 = 2. If just one player becomes a nuisance, he gets 3 + 0 = 3 while the other gets −1. If both refrain, both get 0. These payoffs are all shown in Matrix 4.6.

			Boris	
			Debate	**Refrain**
			(−1, 3)	**(0, 0)**
Horace	**Ping pong**	**(3, −1)**	2, 2	3, −1
	Refrain	**(0, 0)**	−1, 3	0, 0

Matrix 4.6 "Noisy Neighbors." The numbers outside this separable matrix show the separate consequences of each player's decision.

In reality it may not always make sense to simply add up the separate effects of two players' choices as we have done in Matrix 4.6. Thus, it could be that Friday's ping pong game causes Boris to lose sleep, which is not only a loss in itself but also diminishes the pleasure of his Saturday debates by making him slow-witted. The latter effect can occur only if *both* activities take place and one affects the other. The resulting matrix would not be separable since the overall utility would *not* equal the sum of the parts.

If one were given only Matrix 4.6, it would be possible to work backward to separate or decompose the payoffs in that matrix and obtain the numbers in parentheses around the edge of the matrix. Separating payoffs in this way can help to clarify the exact (separate) consequences of a single player's action. Two warnings are in order, however. First of all, not every matrix can be separated; those that can are called *separable* (or decomposable). Second, if a matrix game is separable, there will always be many (in fact, infinitely many) ways to separate it. We will amplify these two points in turn.

For a game to be separable, each player must have a dominant strategy. But this is not the only requirement for separability. Notice that the Noisy Neighbor always does better by making noise, so that noise-making is the dominant strategy. In fact, in a manner of speaking, noise-making is dominant by 3 units; that is, for any choice by the other player, noise-making is exactly 3 units better than refraining. This kind of "dominance by an exact amount" can only be expressed in terms of an interval scale, so it is a more precise kind of property than ordinary dominance, which is expressible in ordinal terms. To determine whether a two-person game is separable, it is sufficient to check that each dominated strategy is dominated by an exact amount.

To see that a separable matrix can be separated in various ways, consider the "Jelly Bean Game," in which you may either give the other player a jelly bean or not. If you do *not,* the Jolly Green Jelly Bean Giver, perhaps to show you the meaning of altruism, gives *you* two jelly beans. The actual jelly bean transactions, shown around the edge of Matrix 4.7, are different from the sepa-

			B	
			Get two from JGJBG (0, 2)	**Give one to other player (1, −1)**
A	**Get two from JGJBG**	**(2, 0)**	2, 2	3, −1
	Give one to other player	**(−1, 1)**	−1, 3	0, 0

Matrix 4.7 "Jelly Bean Game."

rated values around Matrix 4.6, but the resulting combined values in the two matrices are identical. We have thus given two (of the infinitely many possible) ways of separating the values inside the matrix. It is worth noting, though, that for both separated versions as well as for the matrix itself the following statement holds true: a player always gains 3 units when he switches to the dominant strategy and loses 1 unit when the other player does so. Thus these three versions may look different psychologically, but with respect to the interval scale they are equivalent.

A game can be separable, or symmetric, or neither, or both. The example used here has both properties, as does the "Prisoner's Dilemma" in Matrix 4.5. "Chicken," discussed in the next subsection, is symmetric but not separable. In contrast, "Rhodesia" in Section 6.1.1 is separable but not symmetric.

The payoff ordering of Matrix 4.7 (and of Matrix 4.6, which is identical to it) is shown in the next subsection as Matrix 4.8b, where it is given the name "Convergence." The interests of the two players converge but they do not coincide. They do not coincide since each player harms the other by pursuing his own interest. Their interests converge, however, in that if each picks his dominant strategy, they both do better than if both did not. The harm each can do is small enough that it makes sense to allow it to occur. Altruism here would be misguided.

Advice is easy to come by, but it is harder to know which advice to use. Slogans and maxims like the Golden Rule, "Look out for number one," and "Harambee" (Swahili for "Let's all pull together") are evidently in conflict with each other. Perhaps, however, there is a time and place for each. One value of game analysis is that it can help remind us that there are different categories of situations, calling perhaps for different kinds of responses. Thus "Look out for number one" may be appropriate for the "Convergence" game, but for "Prisoner's Dilemma" "pulling together" is more to the point. In Section 4.2.4, you may wish to contemplate the Golden Rule in the context of the game "Spite."

4.2.3 Chicken

"Chicken" is a model of armed confrontation rather than of an arms race. It models the decision whether to *use* weapons rather than whether to build them. The ordinal-scale definition to be given in this section fits with the game of "Nuclear Chicken" in Section 2.3. Here we shall be principally concerned with the strategic properties of the game rather than with analyzing real instances of armed confrontation or finding additional real-world situations that can be translated into a "Chicken" matrix. Of course, anything that is said about a game in the abstract can be considered as possibly relevant to a real-world situation insofar as the game properly abstracts the situation.

To provide perspective on "Chicken," we begin with a brief survey of strategic considerations in various symmetric games. The two symmetric games already introduced and the three still to be examined are shown as Matrices 4.8a–e. The frustration and challenge for the individual in "Prisoner's Dilemma" arises, as we saw in Section 4.2.1, because dominant strategies lead to an outcome that both players could improve upon by joint effort. In Matrix 4.8a this is the lower-right cell containing (T, T), each player's next-to-worst outcome. Dominant strategies are more useful in "Spite," giving the cell with (B, B) each player's best outcome. However, in other games, such as "Chicken" and "Hero," there is no dominant strategy. What can we expect players to do when neither has a dominant strategy?

Failing to find a dominant strategy, a player might settle for his maximin. In both "Chicken" and "Hero" the maximin choice

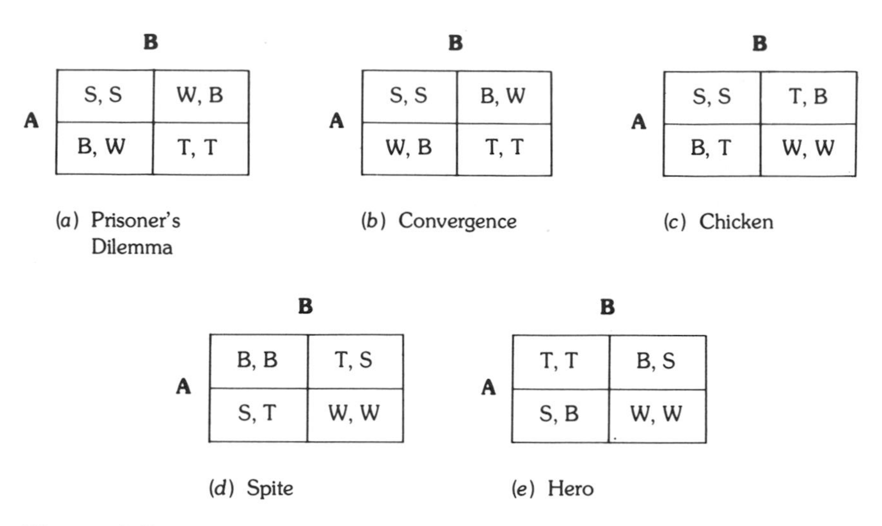

(*a*) Prisoner's Dilemma (*b*) Convergence (*c*) Chicken

(*d*) Spite (*e*) Hero

Matrix 4.8 Five symmetric games.

among pure strategies for player A is the top row, which avoids the worst outcome, W. Looking just at "Chicken," the use of the maximin pure strategy by each player results in the upper-left cell containing (S, S), each player's second-best outcome. This seems a reasonably desirable resolution of the game since, unlike the (T, T) outcome in "Prisoner's Dilemma," which results from the dominance principle *or* the maximin principle, this (S, S) outcome cannot be improved upon jointly by both players.

The similarity between "Chicken" and "Prisoner's Dilemma" has not been apparent so far. One has dominant strategies intersecting in (T, T) while the other has no dominance but has maximin pure strategies intersecting in (S, S). Nevertheless, the two games share an important strategic property, one that tends to create similar motivations for those who play them, whether in laboratory or real-life situations. The relationship between the two games is clearer with interval utility scales than with ordinal ones, so refer to "Prisoner's Dilemma" in Matrix 4.5. In this matrix replace the twos in the lower-right cell by zeros to obtain Matrix 4.9. The ordering of the numbers in this new matrix corresponds exactly to the ordinal entries in Matrix 4.8c, the definition of "Chicken." Thus the alteration of a single cell can convert "Prisoner's Dilemma" into "Chicken," and vice versa.

3, 3	1, 4
4, 1	0, 0

Matrix 4.9 "Chicken," formed from Matrix 4.5 by changing only the bottom-right cell.

To see just how very similar the two games can be, replace the (2, 2) of Matrix 4.5 by (1, 1) instead of (0, 0). The lower-right cell is now neither next-to-worst (for both players), as in "Prisoner's Dilemma," nor worst, as in "Chicken," but *tied* for worst. This borderline game—a sort of "Chickoner's Dilemma"—can be tipped into either of the two game categories by the slightest changes in the (1, 1) cell. Values of (1.01, 1.01) make it "Prisoner's Dilemma" while values of (0.99, 0.99) make it "Chicken." By calling the lower-right cell the "punishment" outcome, one can regard "Chicken" as "Prisoner's Dilemma" with severe punishment.

What can we expect a player in "Chicken" to do? As usual with matrix games, we assume that communication is impossible, so that each player acts without knowing what the other will do. We have already pointed out that, failing to find a dominant strategy, a player might settle for maximin. (Since a player's best two outcomes, B and S, are not on a diagonal, mixed strategies are not relevant to maximin here.) If both players choose their maximin—and since the game is symmetric, there is no reason offhand not to expect the same choice by both—then the result is (S, S) in the upper left (Matrix 4.8c). But there is a difficulty with this conclusion. Suppose, for instance, that you are player B. In line with what has just been said, you anticipate that A will choose the top row, containing the (S, S) outcome. If you are correct about this, then by picking the *right* column you can bring about the payoffs (T, B), giving yourself B, your most preferred result.

It follows that maximin (Rationality Principle 4 in the preceding chapter) is no longer compatible with the assumption that everyone is "rational" (Rationality Principle 2). Thus the two principles that led to our solution for zero-sum games have yielded a group-*ir*rational outcome in "Prisoner's Dilemma" and are not even logically consistent in "Chicken."

Another view of the (S, S) cell is that it is not in equilibrium. An outcome is said to be in equilibrium if no player can unilaterally (single-handedly) improve his own payoff. This is the same definition of equilibrium as that in Section 3.3. To see how it applies in a variable-sum game, consider the lower-right cell (T, T) in Matrix 4.8a. If player A changes to the top row unilaterally, i.e., while player B stays with the right column, then A will have decreased his payoff, changing it from T to W. You should check that a unilateral shift by player B also gives that player outcome W. Thus (T, T) is an equilibrium cell in Matrix 4.8a. A matrix need not have any equilibrium, or it may have just one, or, as we shall soon see, it may have more than one.

If each player in "Chicken" notices that the (S, S) is not in equilibrium and chooses the alternative *not* containing that cell, then the result is (W, W), sometimes called joint disaster, the worst outcome for both. This same mutually unsatisfactory result will also be obtained by a pair of maximax players, those who, perhaps overwhelmed by the strategic complexity of it all, blindly pursue their single best payoff.

Exploring dominance, maximin, and maximax in "Chicken" has not led us to any firm conclusions about the best strategy in that game. Perhaps, however, there might be an equilibrium. Ironically, not one but two equilibrium cells are to be found in Matrix 4.8c, namely (T, B) and (B, T). To check (T, B) for equilibrium, note that if player A shifts unilaterally to the bottom row, he gets W, which is not an improvement; player B certainly cannot engineer an improvement from (T, B), where he already has his best outcome. You should similarly check that (B, T) is also an equilibrium, but note that it *must* be one since the game is symmetric.

Summarizing the deliberations of the player in "Chicken," we find that (1) there is no dominant strategy; (2) maximin strategies intersect at a nonequilibrium cell; (3) if that cell is thought to be the outcome, *each* player is tempted to shift, provided the other does not (unilateral shift yields improvement); (4) if both shift, the outcome is joint disaster; (5) maximax strategies intersect in joint disaster; (6) there are two equilibrium cells; and (7) in each equilibrium cell one player gets T and the other B, although the game as a whole is symmetric.

These several characteristics, taken together, do not seem to lead to any firm conclusions about what a game theorist ought to do or about what typical players are likely to do in "Chicken." Our bag of strategic concepts seems to be running low, but there is a separate compartment to be investigated. In discussing "Prisoner's Dilemma" in Section 4.2.1, we made a distinction between individual and group considerations. Strategies like dominance and maximin have to do with a single decision-maker. The idea of equilibrium, defined in terms of unilateral shifts, also focuses on the individual. But the irony of both "Chicken" and "Prisoner's Dilemma" is that it is so easy to end up with an outcome that is "group-irrational," a property that explicitly does *not* focus on the individual.

The group-irrational outcome, one that is worse for both players than some other outcome, is also called "deficient" or "not Pareto-optimal." In contrast, all other outcomes are "nondeficient" or "Pareto-optimal."* In "Chicken," in particular, the cell with (W, W) is worse for both players than (S, S), so that in principle the

* The expression "Pareto-optimal" is a widespread but somewhat misleading term, since for example the (B, W) and (W, B) cells in "Prisoner's Dilemma" fit the definition.

players would be unanimous in wanting to make a joint shift from one to the other. In the negotiated solutions of Chapter 6, all deficient outcomes are immediately removed from consideration.

An arbitrator for either "Prisoner's Dilemma" or "Chicken" might well determine the (S, S) cell in either as the fair and sensible settlement. It is not deficient, and while neither player is fully satisfied, at least each is equally satisfied (in terms of ordinal payoffs). In neither game is this cell in equilibrium so some sort of enforcement may be necessary. Enforcement, of course, will change the payoffs and then the payoff structure will no longer be that of "Prisoner's Dilemma" or "Chicken." However, in a particular real-life instance of either game, enforcement may possibly solve the actual problem even though the theoretical paradox remains.*

4.2.4 Spite and Hero

"Spite" is a game in which players who seek to keep ahead of each other create a dilemma where there need not be one. The game presents a true opportunity for spite only if payoffs are given in money. Therefore, to analyze the game properly, one must distinguish from the outset between money and utility. In many of the laboratory experiments performed with "Spite," "Chicken," and other games, payoffs have been in real money. The same game is typically played over and over again, usually at least 25 times but often 100 or more times, with payoffs in cents per game that can add up to a few dollars for a one-hour experimental session, depending on the choices made by the players.

The payoff numbers for "Spite" in Matrix 4.10 conform to the ordering in Matrix 4.8d. If these numbers were indeed the payoffs for the two players defined on an interval scale of utility, then Matrix 4.10 would be a particular example of Matrix 4.8d. Each player would then have a dominant strategy, the top row or the left col-

	B	
A	4, 4	1, 3
	3, 1	0, 0

Matrix 4.10 "Spite," in which the payoffs conform to the definition of Matrix 4.8*d*.

* An attempt to resolve the paradox in terms of a more comprehensive theory of rationality has been advanced by Howard (1971). An elementary treatment of this work appears in Brams (1975), and discussion of its significance is summarized by Hamburger (1973b).

umn. The resulting upper-left cell would be best for both players and so the game would be pleasant for the players but a bit dull for the theorist.

Now, however, suppose that Matrix 4.10 represents payoffs in cents rather than utility. Then when the outcome is (4, 4), each player gets 4¢. We might assume that since 4¢ is the largest amount of money, each player will regard this outcome as best, so that we would label it B on an ordinal scale. We *could* make this assumption, but it would be just that—an assumption and nothing more.

On the other hand, consider the cell with payoffs (3, 1). Here player A has to settle for 3¢ but has the satisfaction (if he finds it satisfying) that player B is getting only 1¢. This possible satisfaction might provide motivation for departing from the top row. Since the game is symmetric, the same might hold for B: he might find satisfaction in the (1, 3) outcome, getting 3¢ when he might have had 4¢ but knowing that the other player is getting less. A player who thinks this way is sometimes called a *difference-maximizer.*

A new motive now arises for deviating from the top row or left column. Suppose you are A and you believe that B is a difference-maximizer. Your choice is then between (1, 3) and (0, 0). If you object to the (1, 3) outcome, in which you submit to the malice of the other player, then you may prefer the (0, 0) result. In order to get it, you pick the bottom row. Now notice that your choice is just the same as that of a difference-maximizer! In other words, an outside observer, noticing that you picked the bottom row, could not possibly tell whether you anticipated the left column and were yourself difference-maximizing, or whether, anticipating the right column, you believed yourself to be defending against a difference-maximizer.

Perhaps the simplest view of this game arises in its separable or decomposable form. Each player is offered a penny; if he takes it, the other player gets 3¢ in addition to the 0 or 1¢ he chooses for himself. The (1, 3) and (3, 1) outcomes occur when just one player accepts the penny. The (4, 4) arises when each gets 1 + 3, while (0, 0) occurs when both refuse the penny. In this form it is clear that no matter what the other player does, you can always make yourself a penny richer if you are willing to make the other player 3¢ richer.

Getting back to utilities, what we really mean when we say a player is a difference-maximizer is that his highest utility or most preferred outcome is not (4, 4). Therefore, for such players the money values in Matrix 4.10 do not lead to the ordinal utilities in Matrix 4.8d. If a player regards his 3¢ outcome (with the other getting 1¢) as his best and his 1¢ as his worst, then Matrix 4.10, in cents, corresponds to the preference ranking not of "Spite" but of "Prisoner's Dilemma."

Finally, we come to the game of "Hero," shown earlier as Matrix 4.8e. In this game, if each player tries to get his best payoff then both will get their next to worst. Or, one player may heroically yield and accept his second best, thereby allowing the other player to get his best. Some coordination is needed, since if both yield then both get their worst payoff. This latter outcome would seem highly unlikely if the players are allowed to communicate.

Real situations modeled by "Hero" without its lower-right cell include negotiations that can become stalemated. In Section 5.3 we will discuss a situation in which someone tries to sell something to someone. Since the prospective buyer values the object more than the seller does, both can benefit. Each may try to get the better of the deal and either may yield, not wanting the deal to fall through. "Hero" captures this fundamental aspect of the buyer-seller interaction. However, by restricting players to only two alternatives, it models a less flexible situation. Similar comments apply to other forms of negotiation, including, for example, those between labor and management.

Of course, the real hero game (which really is not "Hero," for reasons that are left for the reader to contemplate) is a game played by several people standing on a dock as someone is about to drown. We assume that the principal aspect of utility is satisfaction that the person will not drown, but there is some disutility in getting wet, so the person who goes to the rescue is the Hero. The more people there are on the dock, the more likely it may seem to each person that someone else will do the job. This reasoning, however, is a "self-undermining prophecy," since the more everyone believes it, the less true it becomes. Such a rationale can lead to the phenomenon of "bystander apathy," as in the infamous case of Kitty Genovese, who was stabbed to death in New York City over the course of many minutes while 38 human beings heard her screams and no one was heroic enough to call the police.*

4.3 Threat Games

Threat and force can find expression even in the smallest of matrices, i.e., 2 × 2. After introducing and comparing these two notions, we will look at the threat structure of two exceedingly sensitive and potentially explosive international situations.

* For discussion of this incident and a report of ensuing research, see Latané and Darley (1969).

Some equilibrium points tend to have greater stability than others. In "Spite," for example, the equilibrium point provides each player with his most favored outcome, a situation that is presumably very stable. Almost as stable should be the equilibrium point in "Convergence." Although players get only their second-best outcome, they not only do worse by shifting unilaterally (by definition of equilibrium), but even if their shift elicits a shift by the other player, the end result is not as good as the equilibrium. Two other kinds of equilibria that are noticeably less stable than these are called *threat-vulnerable* and *force-vulnerable.*

A threat-vulnerable equilibrium appears in Matrix 4.11, a matrix that will crop up again in Section 6.1 as the "Rhodesia" game,

	R	
S	4, 1	2, 3
	2, −2	0, 0

Matrix 4.11 Game with a threat-vulnerable equilibrium.

where it will be used in the analysis of negotiated solutions. This matrix has two dominant strategies that intersect in an equilibrium in the upper-right cell. Although neither player can gain by a unilateral shift from this cell, player S can adopt the following ploy: he notifies R that he is contemplating using the bottom now, thereby giving R a payoff of 0. He also points out that R could do better than this by agreeing to the upper-left cell since 1 is better than 0. In other words, S announces the contingent strategy:

☐ Top row in response to left column.
☐ Bottom row in response to right column.

If R (Receiver of this threat) takes S (Sender) at his word, and if R maximizes within the cells he now believes available to him, then S will succeed in doing better than at equilibrium. Because the threat has the potential to succeed in this way, the equilibrium of this game is said to be threat-vulnerable.

It is important to observe that the threatened action must not be carried out. That is, if S actually did use the bottom row, R would certainly not be likely to switch to the left column. Nevertheless, S must convince R that he *will* carry it out if need be. There is something of a problem here. How do you convince someone you will do something when in fact you won't? Various answers have been suggested, including getting ready to do it and even partially doing it (also see Sections 4.3.1 and 5.3). If the same game or similar ones have been played in the past and you have carried out other threats, your credibility may be high. But looking back at those past acts, the same dilemma applied to them.

A threat will not be very credible if it would be damaging to the threatener himself to carry out the threatened action. For example, a parent may shrink from carrying out a spanking because he sincerely believes the admonition that it would "hurt me more than it hurts you." The child's knowledge that his parent feels this way reduces the credibility of the threat of spanking. It need not be concluded that game theory endorses spanking. One might say that the child's realization that his parents are reluctant to spank is a maturing experience, i.e., may alter the child's utility scale. At the same time, one may well imagine that a child who learns to behave under the threat of punishment may not be learning good behavior but only some rule like "Good behavior is worthwhile when a spanker is around." In this way, perhaps unfortunately, he may get prepared for the enforcement games of Section 4.1. Such a rule might be a precursor of an adult rule, "Obeying the law is worthwhile when a policeman is around." This analogy shows that parent-child interaction is relevant to the authority structures, including policing, of society. More generally, our actions in a real game typically have long-run as well as immediate consequences.

Force-vulnerability, like threat-vulnerability, is potentially destabilizing to an equilibrium point. Unlike the threatening move, however, the forcing move can be effective when actually carried out, and in fact this is the way it does become effective.

● **Game 4.3 Student and Professor**

Professor's first priority is to get Student to do some work, but this takes work on the part of Professor, whose second priority is not to do any work herself. Student's first priority is to do well, which will mean working if work is demanded, but his second priority is not to work.

This verbal description of the key features of a situation can be translated into a game model, in the manner of Section 2.3. In view of Professor's first priority, we deduce that her two most-preferred outcomes are those in which Student works. In view of her second priority, the better of these two top outcomes is the one in which she herself does not work. Her two worst outcomes are those in which Student does not work; of these, her very worst is when she works in vain.

Student's best outcome is to pass without working, which occurs when neither he nor Professor works. Worst for him is to fail, by not working when Professor does. The intermediate results are to pass with work; of these two, presumably the less desirable is to have been fooled. Or we may assume that Student has some interest in learning, so that, for example, if he writes a term paper he at least wants Professor's comments. All of these preferences are summed up as Matrix 4.12.

		Student	
		Work	**Not work**
Professor	**Work**	S, S	W, W
	Not work	B, T	T, B

Matrix 4.12 "Student and Professor," a game with a force-vulnerable equilibrium.

It is left to the interested student, professor, or other reader to determine the following: (a) any dominant strategies; (b) the equilibrium point; (c) the player with a *forcing* move; (d) the forced response of the other player; (e) in what way the new cell, arrived at by each player shifting as in parts *c* and *d*, offers a temptation to the original forcer; (f) why the original forcer will be wise to withstand this temptation; (g) whether the equilibrium point is threat-vulnerable as well as force-vulnerable; and (h) what other situations can be understood in these terms.

4.3.1 Oil-Field Games

> The Arab World replied with a torrent of criticism Tuesday to James Schlesinger's remark that the Arab oil embargo might prompt public demands for military intervention in the Middle East.... Damascus radio warned that the Arabs were fully prepared to blow up their oil wells if force was used against them.
>
> [*Los Angeles Times,* Jan. 9, 1974]

> It's Official—Kuwait's Oil Fields Mined
>
> [Headline, *Los Angeles Times,* Jan. 10, 1974]

The threats in this news story did not occur in a vacuum. The conflict in the Middle East has a long and complicated history, rooted in religion and culture, economics and military strategy. Nevertheless, we shall begin with a very simple tree, incorporating only the possibility of military intervention and blowing up the oil fields. This is because we are more interested in the general notion of threat strategies than in an analysis of the Middle East situation.

Game 4.4 Blow-Up
Saber Rattler, a mighty oil-consuming nation, may invade or refrain from invading Oil Producer, who, if invaded, may blow up his oil fields or not.

The idea of the explosion threat, as shown in Tree 4.1, is that outcome A is worse than E; that is, Saber Rattler is supposed to

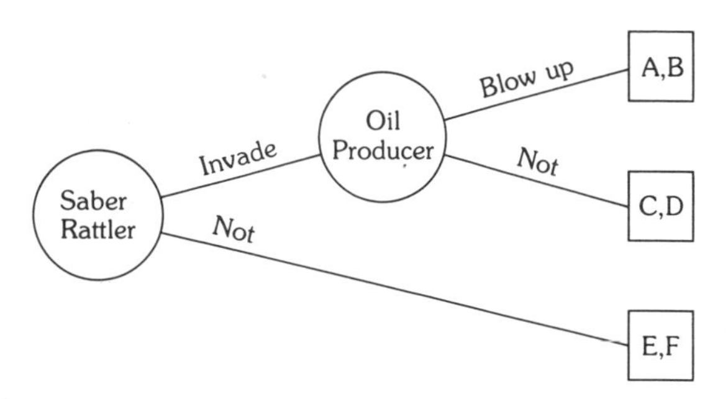

Tree 4.1 "Blow-Up" (payoffs before the comma go to Saber Rattler).

refrain and get E because if he invades he will get A, which is worse. Of course, Saber Rattler will only get A if Oil Producer goes through with the threat, taking payoff B, presumably his worst, for himself.

With the assumptions just made about preferences, namely that A is worse than E and that B is the worst among B, D, and F, the game of "Blow-Up" has a game-theoretic solution. Working backward through the tree, we find that Oil Producer goes last, at the point where invasion has already occurred. In this event, Oil Producer prefers not to blow up his oil fields. It follows that Saber Rattler's decision is not encumbered by any *credible* threat, so that if he prefers C to E he does indeed invade.

It will be interesting to see whether and how this threat tree relates to the notion of threat-vulnerable and force-vulnerable equilibria defined for matrix games in the introduction of this section. To do this, we have to translate Tree 4.1 into a matrix, using the technique of Section 2.4. Each player has just one decision point, each with two alternatives. The result is the 2 × 2 Matrix 4.13a, which becomes Matrix 4.13b in light of the above arguments about who prefers what (B = best; S = second best; W = worst).

Examination of Matrix 4.13b reveals that the upper-right cell is an equilibrium cell and that it is both force-vulnerable and threat-vulnerable. The forcing, threatening move is to "blow up if

(a)

		Oil Producer: Blow up if invaded	Oil Producer: Not
Saber Rattler	Invade	A, B	C, D
	Refrain	E, F	E, F

(b)

		Oil Producer: Blow up if invaded	Oil Producer: Not
Saber Rattler	Invade	W, W	B, S
	Refrain	S, B	S, B

Matrix 4.13 Matrix (*a*) is a translation of Tree 4.1. Matrix (*b*) shows the players' preferences.

invaded." But how can you "blow up if invaded" without knowing whether you are invaded? And after you know, it is too late to affect the other player's decision. Oil Producer needs to bind himself to a contingent plan of action in addition to issuing a warning. He needs a brainless assistant (like the one in Section 2.4) who will carry out the instruction "blow up if invaded." The above headline shows that Kuwait has found an equivalent of an unthinking assistant. The land mines will indeed blow up unthinkingly when there is an invasion. This contingent action would seem to be more effective than a warning that one is "fully prepared to blow up [one's] oil wells if ...." Nevertheless, there remains the problem of how Oil Producer can convince Saber Rattler that the fields are indeed mined. Suppose that Saber Rattler announces his intent to invade tomorrow at noon. Following our analysis, Oil Producer would then turn off the switch, thereby deactivating the mines. But suppose Oil Producer has locked up the switch and publicly blown up the key? But how do they convince Saber Rattler that they do not have a duplicate key? Etc.

Now in the actual case the U.S. did not invade Arab nations in 1974. There are several places to look for an explanation, in this tree and in possible extensions of it. First of all, it may actually be true, as a Damascus broadcast asserted at the time, that "the Arabs have lived long without oil and can and are ready to live long without it. But certainly they cannot live without dignity...." That is, an Arab nation may actually prefer the blow-up to loss of dignity through failure to react (in Tree 4.1 this would mean Oil Producer prefers B to D). Regardless of its true preference, the Arab nation will do well to convince the U.S. that the blow-up has a non-zero probability of being used. One way to do this is for its chief of state to act eccentric and thus be thought "irrational." The U.S. must assess the probability that Syria's dignity is worth more to her than her oil fields. If, for example, it judges that probability to be 1/2, then a decision to invade would yield an expected payoff of

1/2A + 1/2C; this is to be compared with the payoff E in reaching a decision.

Returning to Tree 4.1, note that Saber Rattler may also be misrepresenting his utilities; specifically, he may prefer E to C. Translating this possibility to the real-world example, even if the blow-up threat is not believed, there are many reasons why the U.S. may not really wish to embark on an invasion ("world opinion" and "the lesson of Vietnam" were two reasons mentioned at the time). If this is true, then the question arises as to the real point of the comments by James Schlesinger (then U.S. secretary of defense). An answer to this question requires expansion of the tree. The threat to invade came in response to an oil embargo, prompted in turn by Arab dissatisfaction with previous U.S. acts. Thus, the decision whether to invade or not comes not at the beginning but no sooner than the third stage of this interaction, as Tree 4.2 shows.

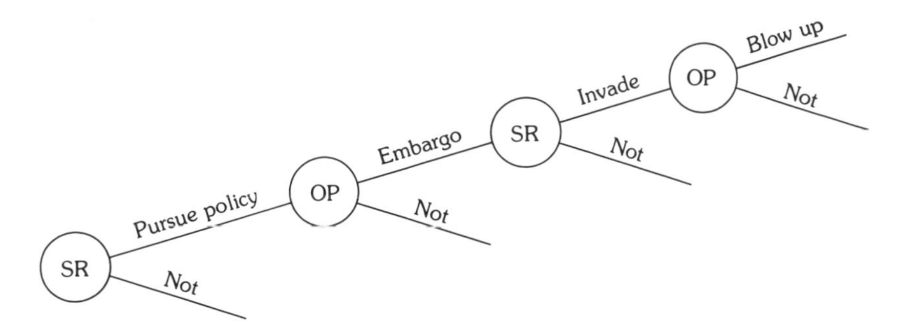

Tree 4.2 Expansion of Tree 4.1. SR = Saber Rattler; OP = Oil Producer.

Focusing now on the embargo-invasion segment in the middle of this larger tree, one may imagine repeated plays of the game as represented by Tree 4.3. The game is played perhaps daily or

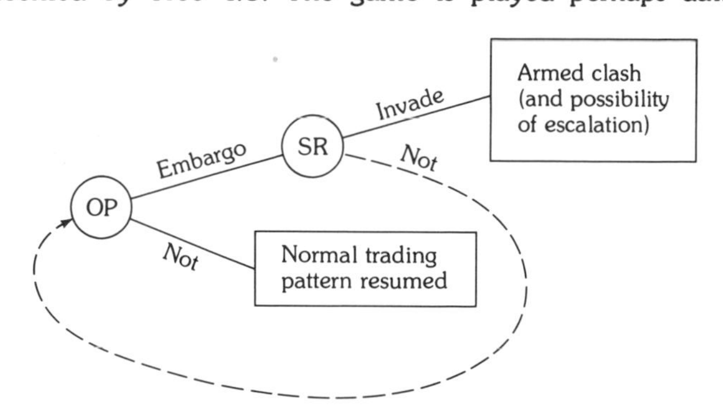

Tree 4.3 A representation of the oil embargo game showing the ongoing nature of the situation.

weekly. The dashed arrow in the tree shows that as long as the U.S. does not invade, the Arabs continue to have the option to continue or to stop the embargo.

In reality, Schlesinger's comments came too late to forestall the starting of the embargo, and one might wonder how they could be believed when many "plays" had already gone by, thus establishing a sequence of embargo followed by no invasion followed in turn by continued embargo. But the successive plays may not be quite exactly repetitions of each other. The U.S. can argue that the utility of no-invasion gets lower as its need for oil gradually increases, saying in effect, "It is true I have not invaded yet, but this does not mean I won't if the situation keeps getting more burdensome for me." There is a certain irony here since an increasing burden is just the effect intended by the embargo, except that the Arabs presumably had a more agreeable U.S. response in mind. On the other hand, the U.S. could claim that the embargo does *not* hurt, and that the oil producers are only harming themselves through lost revenues. But the claim "I can't be hurt" cannot very well be used together with "Don't hurt me too much or I'll strike back."

4.3.2 Balance of Terror

"War," as the Vietnam protest poster declared, "is dangerous to children and other living things." Nor is it cheap. These two aspects of military interaction, money and danger, are roughly modeled by "Prisoner's Dilemma" and "Chicken," respectively. In one you decide whether to build implements of massive destruction, in the other whether to use them.*

The actual military situation in which the United States and the Soviet Union now find themselves is too complex to be modeled by a 2 × 2 matrix. We can, however, gain some insight by regarding the situation as two simultaneous games, one involving decisions about the making of weapons, the other involving decisions about using them. If two nations cannot jointly agree to refrain from building weapons, then, as a kind of lesser evil, they might mutually manage to arrive at opposing collections of weapons systems that do not give either side the prospect of a successful attack. "Successful

---

* See Sections 4.2.1 and 2.3. This characterization of "Chicken" assumes a single decision whether to use massive weapons. When there is instead a succession of decisions, gradually escalating the level of warfare as in Vietnam, and when the implements of destruction are less than massive, then the situation resembles a series of "Prisoner's Dilemmas" in separated form.

attack" here would mean one that overwhelms the other side to such an extent that its response could not possibly inflict major damage on the aggressor.

One such stable situation would consist of both sides having only defensive weapons. For example, the ABM (antiballistic missile) is used to shoot down other missiles. It is a short-range missile that cannot cross oceans and so cannot be used to start an intercontinental war. Unfortunately, as we shall see later, the ABM could indeed *participate* in a "successful attack" along with other weapons. Thus, there is not much point in talking about a single weapon as being offensive or defensive, unless we speak also of a nation's entire arsenal and whether it is capable of a successful first strike against another nation's arsenal.

Perhaps surprisingly, there is a certain ghoulish stability to both sides having only offensive weapons. Consider a world in which ABMs do not exist, and in which two nations have only long-range offensive missiles, lots of them, each carrying very big bombs. Now suppose that Nation A launches its missiles and knocks out Nation B's major cities. Presumably, Nation B will respond with a counter-attack. In fact, it is precisely because A fears B would respond in this way that A, in actuality, refrains from striking first. This is called deterrence: A is deterred from a "first strike" by the knowledge that B's "second strike" will inflict unacceptable damage. Similarly, of course, B is also deterred from a first strike.

Notice that B may have somewhat less fire power than A and still be able to successfully deter A. Thus, it may be possible for either side to disarm slightly without disturbing the balance of terror, that is, without undermining the mutual deterrence. This is important since it means that disarmament need not depend on a negotiated agreement but can conceivably begin spontaneously.

Suppose that Nation A were to make a first strike not against cities but against B's missiles; that is, A's strike is counter-*force* rather than counter-*value*. If A's missiles are absolutely accurate, if B's are all unhidden and unprotected, and if A has more missiles than B, then A can wipe out B and have missiles left over. According to our definitions, this would be a "successful attack," so the situation is not "stable."

What can B do to ameliorate his condition? One possibility is to build more missiles, but presumably A can also build missiles, so this may be futile, not to mention costly. Alternatively, B may try to protect his missiles in any of several ways: "hardening the site" by building protective structures or putting the missiles underground, surrounding the long-range missiles with ABMs, or making the missiles mobile and hard to find by putting them, for example, on submarines.

If these techniques are used sufficiently by both sides, then once again a first strike becomes unattractive since it cannot succeed in wiping out enough missiles to prevent a devastating counter-attack. This makes for a stable situation. When two nations negotiate an arms agreement, that is, when they together arrange the rules of the game they will then both play, their mutual objective is to create a stable situation like the one just described. At one point in the cold war the United States actually was encouraging the Soviet Union to harden its missile sites. The argument ran something like this: "If you don't harden them, you'll be worried that we might feel we could attack successfully (first strike) and such a concern may lead you, in a moment of world tension, to get desperate and try to get off a first strike of your own, which would not be successful, but would be disastrous for both sides."

It is one of the peculiar ironies of strategic thinking that, at least in one view, ABMs turn out to look more threatening when emplaced around cities instead of missile sites. At first glance nothing could look less offensive than short-range missiles arranged to protect a civilian population. The trouble is that city-ABMs would be only a part of a nation's weapons system. Suppose both A and B have lots of long-range missiles but A has more. Further imagine that missiles are fairly accurate but sometimes miss.* Suppose A can knock out most of B's missiles but that B's weakened force could still inflict unacceptable damage on A in a counter-attack. Then B has a deterrent and the situation is stable. Into this situation let us introduce ABMs around A's cities. Now the situation is much less stable, since an attack by A is more likely to "succeed." This is so because a counter-attack by B's weakened force, directed against A's cities, can be sufficiently blunted by A's ABMs to prevent unacceptable damage.

An important aspect of the foregoing scenario is the accuracy of the first-strike missiles. If they are extremely accurate, then the counter-attack will be weak indeed, so that a first strike is tempting. The situation may thus be destabilized even in the absence of ABMs. This makes it clear why the idea of a cruise missile, designed to fly below radar range, carry its own computer-based maps, and correct its own path, is such a hot topic at SALT II (see "Cruise Missiles" by Kosta Tsipis in *Scientific American,* February 1977).

* "Missing" may be interpreted as not coming close enough, where "close enough" depends on how much the missile site has been hardened. Also, if A has extra missiles, it may aim more than one at a missile of B, thereby increasing the probability of a hit.

It might seem that the best protection of the viable counter-attack, and hence of stability, is the submarine-launched long-range missile. This scheme has the drawback, however, that it effectively distributes the ultimate power to start a war into the hands of many submarine commanders (see "The Failure of Fail-Safe" by John Raser in *Trans-Action,* January 1969).

In summary, game theory can distinguish three aspects of military decision-making: negotiating the rules of the game; the building of armaments; and attack and counter-attack. The foremost objective is to prevent any attack. However, a simple rule directly disallowing attacks would be futile, because nations would not trust each other to follow such a rule. Therefore, the rule-writers attempt to prevent attack indirectly through arms control agreements. They seek rules under which only stable situations can emerge, those in which attacks will not look attractive to either side. Other key objectives for rule-writing are to discourage expenditure and to avoid accidental attack.

Exercises

○ **1** For the matrix shown,

1, 4	−2, 3
3, 2	−5, −2

(a) Find any equilibria.

(b) Find Row-chooser's minimum or worst possible payoff in each row and write it to the right of the row.

(c) Which of the row minima in part (b) is the best or maximum of the minima (maximin)? Which row is it in?

(d) Find Column-chooser's minimum in each column.

(e) State the maximin value and maximin strategy for Column-chooser.

(f) If both players use maximin, what cell is the result? Is this an equilibrium?

○ **2** "In the 77 episodes from 1968 to 1975 in which hostages were held for ransom, the victims included about 30 American officials, six of whom were killed" (J. Miller, "Bargain with Terrorists," *New York Times Magazine,* July 18, 1976). Consider the following three statements discussed in the article cited: (1) Refusing to negotiate, pay ransom, or make political concessions deters terrorists from kidnapping American officials. (2) Terrorists want a lot of people listening, not a lot of people dead. (3) What really deters is not a hard line during the crisis, but determined action afterward to capture and convict the terrorists.

(a) Incorporate some of these ideas into a simple tree or matrix of the situation.

(b) Comment on the relative payoffs associated with various outcomes and on the resulting strategic considerations.

○ **3** Think of other situations in which one party tries to get the other to do something using various tools of inducement. Use game-theoretic notions in discussing whether your examples are more like the speeding or the terrorist situation.

○ **4** In a separable, symmetric game each player (call them A and B) must choose exactly *one* of the following: either (I) receive 2¢ while the other player receives 0, or (II) receive 0 while the other player receives 4¢.

(a) Give the matrix of this game *and* state which of the five game types of Section 4.2 it exemplifies.

(b) Same as part (a) but with alternative II replaced by: (II′) receive 0 while the other player receives 1¢.

(c) Same as part (a) but with alternative II replaced by: (II″) receive 0 while the other player loses 5¢.

(d) Can there be a separable game that has the payoff ordering of "Hero"? If yes, give one. If not, why not?

○ **5** **(a)** Explain why the following game is symmetric. (You may wish to relabel rows or columns.)

		Coretta	
		i	**ii**
Bella	**I**	S, W	T, T
	II	B, B	W, S

(b) In a certain matrix game Row-chooser strongly prefers that Column-chooser select the left column and weakly prefers that he himself choose the top row. His preferences among cells of the matrix are determined simply by combining the effects of the corresponding choices. Draw a matrix that shows appropriate ordinal payoffs (B, S, T, W) for Row-chooser. (Leave Column-chooser's payoffs blank.)

○ **6** In the discussion of Game 4.3, "Student and Professor," eight questions are posed. Answer the first seven of them.

○ **7** The cruise missile is very good both at finding its target and at not being shot down while doing so. How could any right-thinking citizen not want her country to have one?

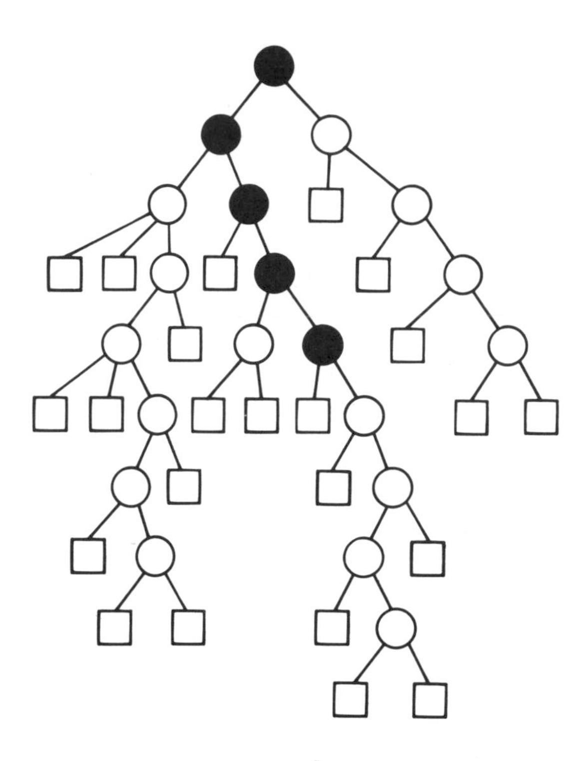

5 Circumstances of Play

Most of the games up to this point have been 2×2 games, though a few have presented three or more options. In contrast, many situations present a virtually unlimited number of options. A potential candidate for office can make his announcement at any hour of any day over a period of months. If a city budget can be \$50 million, then it could probably also be \$51, \$52, or \$49 million. If a game is repeated many times, then a series of decisions must be made. If these are assembled as a single, complex decision, then again there are a huge number of options from which to choose. These three kinds of options—timing, gradation, and repetition—are discussed in the first two sections.

A payoff structure reveals something, but not everything, about a situation. It may show, for example, that a game has potential for threat by one player, but it does not tell whether in fact the player has available the means to communicate that threat. Moreover, the very act of writing down the payoff structure itself may be deceptive in that it suggests that the payoffs are known. Perhaps you as analyst of a situation feel you know the preferences of players, but that is no guarantee that the players themselves can gauge each other's utilities. In fact, it is sometimes in a player's interest to deliberately misrepresent his utilities. Considerations of this sort make up the third and final section of this chapter.

5.1 Shades of Gray

Some decisions are black-and-white, either-or propositions. If you are deciding whether to quit your job, then either you quit or not—you cannot "half-quit." However, black-and-white decisions can be blurred into shades of gray in a variety of ways, and many situations are inherently gray. Gradations are one type of intermediate course. For example, in our description of an arms race in Chapter 4 the options to each player were "high" or "low" arms budget, but in reality a nation can raise or lower its arms budget in graded amounts—by one or two billion or more as it sees fit. This topic is pursued in Section 5.1.2. In some black-and-white situations gray strategies can be introduced by mixing options over repeated playings of the game. Thus, on a given day I must decide whether to drive to work or take the bus—a black-and-white decision in that I cannot do a little of each. But in the course of a week or a month I can indeed pursue a "gray" course of action, riding on some days while driving on others. A candidate's black-and-white decision of whether to run at all is followed by a gray decision in a game of timing her announcement. A timing decision may also face legislators announcing support for an issue, or, as we shall now see, convention delegates announcing support for a candidate.

5.1.1 Timing

For Ted T. Barr, sheriff of Cabell County, W. Va., the day dawned with the agony of sleepless indecision, but ended with the euphoria of finding himself hailed as a kingmaker....

Barr came to the Republican National Convention uncommitted, one of the handful of delegates who would decide the party's presidential nomination....

On Tuesday, Barr joined two other uncommitted West Virginia delegates in announcing that they would vote for Ford

> on the convention's first ballot tonight. According to the Associated Press delegate count, the decision lifted the President past the 1,130 mark required for nomination.
>
> [*Los Angeles Times,* Aug. 18, 1976]

The agony and the ecstasy experienced by Sheriff Barr were not available to every delegate to the 1976 Republican convention. Most of them could not become celebrated uncommitted delegates because they were elected specifically on the basis of their commitment to one of the candidates, having run as a delegate for either President Ford or Ronald Reagan. Among those who were originally uncommitted, some made up their minds early, either on principle or because of pressure or explanation from one side. For these the utility of helping one particular candidate by publicly endorsing him was sufficient to outweigh any potential reward for remaining uncommitted.

What are the rewards for remaining uncommitted? One is the slim chance of becoming the kingmaker, the one who makes the difference. But the momentary euphoria of Ted Barr is not the only possible payoff. Two months earlier another uncommitted delegate, Bobby Shelton of South Carolina, made news. Shelton needed some political help from Governor James Edwards, Reagan's most influential supporter in South Carolina, and he was using the circumstances to his advantage.

> "At one point, I wanted him to come down here for a fund raiser so we could raise some money to recruit local candidates. You know, if you'll run for County Council, the party'll give your campaign $500 or $750 or something like that. I couldn't get him up here for that.
>
> "Now, though," Shelton said, "now he's started to call. I told him last week, 'It's surprising how you just get interested in me now.'"
>
> All of the Republican brass in South Carolina have called him repeatedly, he said, and he refuses to budge.
>
> [*Washington Post,* June 20, 1976]

The uncommitted delegates to the Republican convention of 1976 were engaged in a game of timing. Each of them had to decide when to make a public commitment, so choosing a strategy meant choosing a day. This particular game of timing lasted a long time since some uncertainty persisted almost up to the convention, though in fact by the time Barr announced there was little doubt that Ford would win.

Bobby Shelton's decision problem is indeed one of timing; to take advantage of his situation he must wait, but not wait too long. His bargaining leverage against Governor Edwards increases for a while but then it will decrease. The turning point is determined by all the other delegates, each of whom faces the same options. All of them may be trying to outguess each other. Of course, various ones may have political situations that differ in detail, but each of them wants to be in the front seat of the bandwagon, not flailing after the back rail.

Another political timing game involves candidates announcing their candidacy. Since the players are candidates, rather than voters or delegates, the number of players is much smaller. The decision of when to announce that one is a candidate for President or some other office is one that seems to intrigue the press greatly. In fact, a potential candidate's ability to turn aside reporters' probes on this subject sometimes seems to affect whether she can in fact hold off her announcement. This situation would be particularly intriguing if the following reasonable description were accurate.

Game 5.1 Announcement of Candidacy

Candidates may announce at any time. Payoffs are determined by the following disadvantages:

1 A candidate who announces just after another is disadvantaged (overshadowing effect).

2 A candidate who announces too early is at a disadvantage (boredom effect).

3 A candidate who announces too late is at a disadvantage (disorganization and underexposure).

Disadvantages 2 and 3 suggest that players will select the "middle" period—but exactly *when* within that period? Suppose you expect your opponent to announce in the *middle* of the middle period. Then in view of disadvantage 1 you should announce just before then (still in the middle period, however). But of course if *he* thought *you* would do that, he would try to beat you to it. This line of reasoning, if continued, takes us further and further back toward the early period, which, by disadvantage 2, is not an advantage at all. The trick would be to figure out just when the "middle" period begins, except, of course, that "early" blends into "middle" with no clear dividing line.

The foregoing considerations implicitly assume that the candidates are on an equal footing. Typically, however, decisions on timing are affected also by how well-known a candidate is at the outset. A George McGovern or a Jimmy Carter, virtually unknown nationally at the outset, will start early to gain exposure while the overexposed Hubert Humphrey delayed his 1976 announcement so long that he turned out not to make one.

In another arena, labor negotiations can turn into a game of timing if both sides become rigid in their positions. In the Los Angeles bus strike of 1976 the opposing sides were within half a percent of agreeing on the size of the annual salary raise for a three-year pact, and then weeks went by with neither side giving an inch. Here the trick is to hold out longer than the other side, but of course both sides pay a heavy penalty for each day the strike continues.

5.1.2 Gradation

An arms race, as already noted, allows for gradations in choice of budget level. There is no obvious limit to how fine these gradations may be, but the four levels of Matrix 5.1 will be adequate for the

	$80B	**$90B**	**$100B**	**$110B**
$ 80B	0, 0	−2, 1	−4, 2	−6, 3
$ 90B	1, −2	−1, −1	−3, 0	−5, 1
$100B	2, −4	0, −3	−2, −2	−4, −1
$110B	3, −6	1, −5	−1, −4	−3, −3

Matrix 5.1 "Arms Race with Gradation."

purpose of illustration. As in Section 4.2.1, certain assumptions about utilities are needed in order to translate the situation into a matrix.

● **Game 5.2 Arms Race with Gradation (Guns or Butter)**
Each player chooses a level of military expenditure. Utilities are determined entirely by the following considerations.

1 A player receives two units of utility for every $10 billion by which his budget is above the other player's, or loses two units for each $10 billion under (the guns effect, i.e., a diplo-military advantage).

2 A player loses one unit of utility for every $10 billion he spends beyond $80 billion (the butter effect, i.e., loss of other goods and services).

The choice of $80 billion as a baseline is arbitrary; we could just as well have *added* one unit for each $10 billion of savings relative to a $110 billion baseline. Changing the assumptions in this way would merely add three units to each player in every cell, an inconsequential change from the standpoint of an interval scale of utilities. Incidentally, the use of an interval scale here instead of the ordinal, as in Section 4.2.1, is a matter of convenience.

There are several ways in which Matrix 5.1 deserves to be called "Prisoner's Dilemma with Gradations." The four corners, corresponding to the extreme choices of $80 billion and $110 billion, form Matrix 5.2a, which is "Prisoner's Dilemma" since it fits the

	$80B	**$110B**
$ 80B	0, 0	−6, 3
$110B	3, −6	−3, −3

	$80B	**$90B**
$ 90B	1, −2	−1, −1
$100B	2, −4	0, −3

Matrix 5.2 Submatrices of Matrix 5.1, in which players are restricted to a subset of the options.

definition of Matrix 4.6a. Thus, one could say that Matrix 5.1 is "globally" a game of "Prisoner's Dilemma" in that its extreme options form that game. It is also "locally" the same game in the sense that any 2×2 matrix formed by restricting the options to two adjacent rows and two adjacent columns will also be "Prisoner's Dilemma." For example, Matrix 5.2b has the appropriate payoff orderings. Finally, note that even the original 4×4 game has the key characteristics of "Prisoner's Dilemma," since the $110 billion choices are dominant and intersect in a deficient outcome.

Matrix 5.1 is separable as a direct consequence of the two assumptions in Game 5.2 that were used to construct it. The decision to add an additional $10 billion to the arms budget has two effects: by statement 1 of Game 5.2 there will be a gain of two units with a consequent loss of two by the other player, and by statement 2 a loss of one unit. The net result is a gain of one for oneself, which costs the other two. If both do this, each has a net payoff change of $1 + (-2) = -1$.

Gradations in games have received some attention because of the notion that disarmament might be feasible in step-by-step fashion. The other side of this coin is escalation of hostilities, a phenomenon that can arise in interpersonal as well as international affairs. A final example of a "Prisoner's Dilemma with Gradations" is the price war in which each tiny decrease in price by one company attracts a certain number of its competitors' clients, but the gain to oneself, being modified by one's lower price, is less than the loss felt by the other player.

5.2 Repeated Play

In a tree (but not in a matrix) players may have more than one move. In a matrix (but not in a tree) players move simultaneously. Real-life situations can incorporate both conditions. That is, the parties involved make decisions without communicating, each learns what the other(s) did, and then they make separate decisions again. Play of this sort may be repeated any number of times. And it may stop for any of several reasons. Different "stopping rules" can have quite different effects on repeated games, and we shall compare some possibilities in Section 5.2.1. Then in Section 5.2.2 we investigate whether repetition can provide a way out of the dilemma of "Prisoner's Dilemma."

5.2.1 When the Repetition Stops

What determines how many playings of a game will take place in real-life situations? It is conceivable that the players themselves control this aspect by means of their own actions within the game, according to certain "termination rules." Thus, in labor-management negotiations to end a strike, "play" ends when agreement is reached. To be slightly more specific, let us imagine successive days of negotiation as repeated plays of the following game.

● **Game 5.3 Strike**

Each side has three alternatives: Yield *(give in completely and accept the other side's original proposal);* Split *(offer to accept a wage level halfway between the two sides' original proposals); and* Hold out *(insist on one's own original proposal). If on some day one side yields or both split then the game is played no more, but otherwise the players must return the following day and play the same game again.*

The conditions on stopping or playing again are shown in Matrix 5.3. Note that to be complete this matrix should also include payoffs in every cell. When one player chooses H (hold out) and the

	Yield	**Split**	**Hold out**
Yield	Stop	Stop	Stop
Split	Stop	Stop	Play again
Hold out	Stop	Play again	Play again

Matrix 5.3 Matrix of termination rules for negotiation game.

other Y (yield) the H-chooser gets a high payoff and the Y-chooser a low one. This aspect of the game is like "Prisoner's Dilemma" with Y as the cooperative or dominated strategy. If both yield we can imagine that they then agree on a split. Thus the (Y, Y) outcome gives intermediate payoff values, like joint cooperation in "Prisoner's Dilemma." The (H, H) outcome prescribes that the game be played again but not without penalty. The loss of a day's wages for the union and of a day's production for management must be reflected in a negative payoff. Such negative payoffs accumulate as long as players continue to have the outcome (H, H) or (H, S) or (S, H). These losses must ultimately be charged for each player against whatever gains that player gets from the final settlement.

The game of "Nuclear Chicken" (Sections 2.3 and 4.2) has a termination rule that differs in an interesting way from the one just discussed for "Strike." In both games play may be stopped as a result of certain combinations of choices, but the particular choices that result in stopping are quite the opposite. If one takes the worst outcome for both players in "Nuclear Chicken" to represent nuclear holocaust, then that surely constitutes a termination of the game. Thus, while the relatively more cooperative portions of the "Strike" matrix call for termination, it is joint noncooperation that terminates "Nuclear Chicken."

In order for the foregoing comments to make sense, we must specify in what way "Nuclear Chicken" is played repeatedly. The most straightforward approach is to say that each potential confrontation constitutes a playing of the game. The clearest example of such a playing is the Cuban missile crisis of 1962, where both sides appeared to be actively contemplating D or some kind of "D-like" intermediate strategy like the "split" alternative in the game of "Strike." Other examples with different results might include the Soviet Union's invasion of Czechoslovakia in 1968 and the United States' suppression of the Dominican Republic in 1965. These actions may have been D choices or they may have resulted from intermediate strategies that met with C responses. That is, since in each case the other major power made no significant response, we cannot tell whether the aggressor would have been willing to risk all-out war.

None of the real-life episodes just cited as examples of playings of "Nuclear Chicken" resulted in joint cooperation. A playing that had this outcome would not make the headlines and so would tend to be "invisible," like the salary raise not asked for in the game of Tree 2.8. This is because if both sides yield completely on some potential point of conflict, then it never appears as an actual point of conflict. Matrix 5.4 represents the termination rules and a description of outcomes for "Nuclear Chicken" according to the above view.

		S.U. C	S.U. D
U.S.	C	No payoff (invisible payoff) *Play again*	Diplo-military advantage to USSR *Play again*
U.S.	D	Diplo-military advantage to U.S. *Play again*	Holocaust Stop

Matrix 5.4 "Nuclear Chicken." Matrix shows both the real-world outcomes and the termination rules.

5.2.2 Supergame Strategy

Interest in repeated play originally grew out of the study of arms races as instances of "Prisoner's Dilemma." Arms races are of such great concern that even if they alone were modeled by "Prisoner's Dilemma," that game would be worth serious study. In fact, "Prisoner's Dilemma" is structurally similar to a wide variety of other games: pollution by people, businesses, and nations; contributions of money, time, and blood; and all the other large-scale social and economic dilemmas described in Section 7.2.

This seems to be a very unhappy state of affairs. A game that seems diabolically contrived to elicit unsatisfactory outcomes crops up practically at every turn. Thus, it is natural that game theorists have been at considerable pains to find a way out of the deficient equilibrium of "Prisoner's Dilemma," to devise a theoretical framework within which cooperation will appear as an individually rational alternative. With this purpose in mind, we now turn to the question of repeated play. How do we represent it and what effects will it have on players' courses of action? We will find that a strict logical treatment of these questions gives a rather gloomy conclusion. Interestingly, however, real human beings are not hamstrung by these considerations and, as will be seen in Chapter 9, in experimental games they find ways of reaching cooperative results.

For simplicity, we will take the smallest possible "Prisoner's Dilemma," a 2×2 matrix, and assume that this game is played only two times. Although most of the real-life examples of "Prisoner's Dilemma" mentioned above have large numbers of players, some may have as few as two. Also, in most cases more than two alternatives are available; when your friendly senator asks for $25 for his presidential campaign, he will settle for $20, or $15, or.... Nevertheless, the two-person, two-alternative form captures the essence of the dilemma.

What self-oriented reason can one possibly give for the cooperative or dominated choice or, as it is often called, the C choice in "Prisoner's Dilemma"? Since it is dominated, it offers no possible utility benefit in a single playing of the game. If the game is to be played more than once, however, a player might imagine that cooperation on the first playing may elicit cooperation by the other player in the second playing. There is some reason to think this might work. The C strategy is, indeed, cooperative in that it helps the other player no matter what he may do. This sounds a little like dominance, but it is different. As an example, consider Matrix 5.5. Here, if

		B: C	B: D
A	C	1, 1	−2, 2
	D	2, −2	−1, −1

Matrix 5.5 "Prisoner's Dilemma."

player A chooses the top row, then he helps player B no matter what B may do. If B picks the left column, then the top row is helpful in that it gives B a payoff of 1 rather than the −2 that B would get if A chose differently. Similarly B prefers 2 to −1.

Just as a dominant strategy is self-beneficial no matter what, so we could invent a name for a strategy that is beneficial to the other player no matter what. Let us call it "helpful." So C, although it is a dominated strategy, is also helpful in our sense and may well gratify the other player. Incidentally, a strategy can be dominated without being helpful, or vice versa. For example, the game of "Chicken" in Matrix 4.9, repeated here as Matrix 5.6, has a helpful strategy that is

3, 3	1, 4
4, 1	0, 0

Matrix 5.6 "Chicken."

not dominated; in fact, the game has no dominated or dominant strategy.

It is now time to take a formal look at the precise form taken by "Prisoner's Dilemma" played twice. Although the players move simultaneously from the viewpoint of an outside observer, there is a sequence experienced by each player in which he *learns* about the other's first decision *after* making his own first decision (even though the two decisions were simultaneous). The sequence each experiences is:

☐ **1** Make my own decision in first playing of the game.
☐ **2** Be told what the other player did in first playing.
☐ **3** Make my own decision in second playing.

This sequence for player A is represented as Tree 5.1, in which the branching points are numbered 1–7 for future reference; a similar tree could be drawn for player B. (At this point a rereading of Section 2.4 may be helpful since we are once again going to convert a tree into a matrix.)

Recall that a strategy is a complete contingency plan, a statement that tells what to do in any situation that could arise. It is a statement that could be left to a faithful assistant of very little brain while the player herself goes on vacation. A strategy for A in Tree 5.1

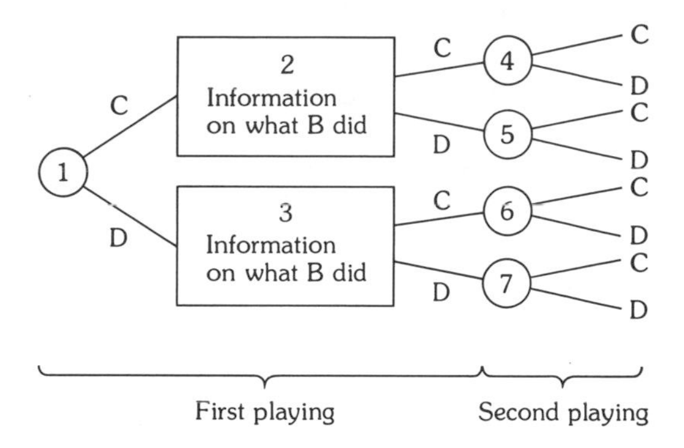

Tree 5.1 Game tree for two playings of "Prisoner's Dilemma." The tree shows the sequence of moves for player A only. Decision points for player A are the numbered circles, referred to in the text. Receiving information does not constitute a decision but may affect subsequent decisions. A similar tree would represent the sequence for player B.

must tell what to do at the start, i.e., at decision point 1. If A picks C at decision point 1, she will end up at decision point 4 or 5, depending on information about B's decision at point 2. Therefore, decisions must be stated in advance for both of these. The possible strategies that include "C at 1" are:

☐ **I** C at 1, C at 4, C at 5
☐ **II** C at 1, C at 4, D at 5
☐ **III** C at 1, D at 4, C at 5
☐ **IV** C at 1, D at 4, D at 5

If, on the other hand, D is picked at decision point 1, then after receiving information about B's choice at point 3, player A will find herself at decision point 6 or 7. The additional strategies are:

☐ **V** D at 1, C at 6, C at 7
☐ **VI** D at 1, C at 6, D at 7
☐ **VII** D at 1, D at 6, C at 7
☐ **VIII** D at 1, D at 6, D at 7

Strategy I is to play C in both the first and second playings, no matter what the other player does. Strategy II, however, is more sophisticated; the second decision is based on the information acquired about the other player's first decision. It calls for C at decision point 4 in response to C, but D at decision point 5 in response to D. Thus, the player using strategy II will do on the second playing whatever the other did on the first—she will give "tit-for-tat," or "a taste of your own medicine." Notice now that strategy VI also calls for tit-for-tat on the second playing, but differs from II in that it uses D on the first playing.

Strategies III and VII are included for completeness, but it is hard to imagine why anyone would want to use them. Each calls for responding with the opposite of the other player's first decision. With all these comments in mind, it is now possible to give a more meaningful description of the eight strategies:

☐ **I** C, then C
☐ **II** C, then tit-for-tat
☐ **III** C, then opposite
☐ **IV** C, then D
☐ **V** D, then C
☐ **VI** D, then tit-for-tat
☐ **VII** D, then opposite
☐ **VIII** D, then D

In this new form no reference is made to the name of the player, and since this is a symmetric situation these can surely be B's strategies as well as A's. Thus, two playings of the game can be represented as an 8×8 matrix, and this has been done in Matrix 5.7.

To see how we fill the cells in this matrix of a twice-played "Prisoner's Dilemma," consider a few examples. If both play "C, then C" (strategy I), then there will be a (C, C) outcome both times. Each will get a payoff of 1 in each playing for a total of 2. Thus, the upper-left cell of Matrix 5.7 contains (2, 2). As another example, suppose player A picks "C, then D" (strategy IV), while player B picks "D, then tit-for-tat" (strategy VI). Player B's tit-for-tat will trans-

	I	II	III	IV	V	VI	VII	VIII
I	2, 2	2, 2	−1, 3	−1, 3	−1, 3	−1, 3	−4, 4	−4, 4
II	2, 2	2, 2	−1, 3	−1, 3	0, 0	0, 0	−3, 1	−3, 1
III	3, −1	3, −1	0, 0	0, 0	−1, 3	−1, 3	−4, 4	−4, 4
IV	3, −1	3, −1	0, 0	0, 0	0, 0	0, 0	−3, 1	−3, 1
V	3, −1	0, 0	3, −1	0, 0	0, 0	−3, 1	0, 0	−3, 1
VI	3, −1	0, 0	3, −1	0, 0	1, −3	−2, −2	1, −3	−2, −2
VII	4, −4	1, −3	4, −4	1, −3	0, 0	−3, 1	0, 0	−3, 1
VIII	4, −4	1, −3	4, −4	1, −3	1, −3	−2, −2	1, −3	−2, −2

Matrix 5.7 A matrix translation of Tree 5.1.

late into a C since player A has picked C on the first playing. The resulting computation of total payoff is as follows:

	First playing		**Second playing**		**Total payoff for the two playings**
	Decision	Payoff	Decision	Payoff	
Player A	C	−2	D	+2	0
Player B	D	+2	C	−2	0

Thus, (O, O) appears in row IV, column VI of Matrix 5.7. The other cells are computed similarly.

The game in Matrix 5.7 is called a "supergame" because it is one game but is formed from more than one (the two playings). What insight does the supergame provide into the strategic impact of repeated play? To answer this question let us look for dominated strategies. It turns out that six of each player's strategies are dominated. In particular, IV dominates I, II, and III, while VIII dominates the other three. Removing dominated strategies in the manner of Section 3.3, we are left with the residual game of Matrix 5.8. In this

	IV	VIII
IV	0, 0	−3, 1
VIII	1, −3	−2, −2

Matrix 5.8 Refinement of Matrix 5.7, removing all dominated strategies.

game strategy VIII is dominant for each player. But strategy VIII is nothing but "D, then D," the repeated use of D. Thus, our analysis of repeated play merely confirms the apparent inevitability of the D alternative. Some mitigation of this judgment may lie in the fact that "D, then D" is not dominant in the full supergame of Matrix 5.7, but only in its reduction in Matrix 5.8. Strategy VIII is thus rational only if one assumes that the other can find and eliminate the dominated strategies of Matrix 5.7. Moreover, as you may have noticed, the computation is fairly complex. This is significant, since we may not want to say someone is irrational just because he does not spontaneously come up with this rather complex line of thinking. So repeating the game does not succeed in making anything other than repeated use of D individually rational, but it does show us that this is individually rational only if one assumes that the other player is "rational" in a fairly sophisticated way.

In a way it is not surprising that the strategy "D, then D" turns out to have a strong rationale for a payoff-maximizing player. After all, there is no reason to use C on the second playing of the game—it could not possibly induce a cooperative response since play stops after the second playing. Thus, from the standpoint of "individual rationality," it is a foregone conclusion that D will be used on the second playing, and this in turn makes it futile to use C on the first playing in hopes of eliciting a C response on the second. A little reflection reveals that in three playings of "Prisoner's Dilemma" similar considerations prevail: D is a certainty on the third playing, hence C is futile on the second, so D is also a certainty on the second, making C futile in the first. This reasoning can be extended to any number of playings as long as the exact number is known in advance.

However, the exact number of playings certainly might not be known. In various real-life situations a conflict may elicit a series of decisions and then, at some point, subside or become unimportant for unforeseen reasons. Players typically do not have an exact notion of how long a conflict will last. To make the idea of uncertain length clearer, let us take a specific example. Suppose that a game of "Prisoner's Dilemma" is to be played two or three times, but the players do not know which. Is repeated D still rational in the same way as before?

To answer this question it is first necessary to ask what a strategy will look like in a game played two or three times. Is it even possible to specify a complete plan of contingent decisions if we do not know how long the game will go on? The answer to this question is yes. All we have to do is specify a strategy for three playings; then, if the play stops early, our specifications for the third game simply go unused. But, you may ask, suppose one wants to play differently depending on whether the game is to be played two or three times? The answer is simply that this cannot be done; it is impossible to base decisions on something unknown. Thus, a strategy for two- or three-shot "Prisoner's Dilemma" is simply a strategy for three-shot "Prisoner's Dilemma." Once again, any "individually rational" strategy must call for D on the third playing. For the second playing, the choice of C is futile since even if a third playing did take place, the choices in it would not, by our previous reasoning, be influenced by choosing C. Thus C is again futile even in the first playing, since again all future choices are determined.

This is the "gloomy conclusion" promised earlier. Many experimental subjects who play repeated "Prisoner's Dilemma" for real money are able to overcome, or remain blithely unaware of, the foregoing arguments. The behavior of players in experimental games will be discussed in Chapter 9, but here we may note that the tit-for-tat strategy introduced above seems to be the key to eliciting cooperation from another player, at least in two-person "Prisoner's Dilemma."

5.3 Information, Communication, and Bargaining Stratagems

Information comes in many varieties. You may have some kinds but not others. Some kinds of information may be very important to you, others unimportant, and still other kinds you may be better off without. For example, you may want to know when the bus is coming, not care where it is coming from, and, if you have some reading you want to get done during the ride, want *not* to know which of your talkative acquaintances is sitting in the rear, as you slink, eyes down, into a front seat.

A person, nation, party, or corporation may have or not have, want or not want, information about itself or others, information about the tree structure or payoffs, or about what the other players are doing. Moreover, a person, nation, party, or corporation may have or not have information about information, i.e., information about the other player's state of information. And there may be or may not be the possibility for players to exchange information, that is, to communicate. If they can communicate, they can transmit threats, warnings, promises, information, or even misinformation, and they may be able to negotiate a coordinated set of actions that benefits both (or all) of them.

Information and communication capabilities may be symmetrically distributed. Alternatively, one player may have more information than another or may be better equipped to receive or to transmit messages. Experimental studies (see Chapter 9) have shown that players of "Prisoner's Dilemma" tend to cooperate more when both of them know the payoffs than when both do not, and choose the cooperative C option much more if they can communicate with each other. Such results may be taken as one rationale for the Washington–Moscow hotline and, more generally, for full and symmetric information and communication. On the other hand, in a situation where one player has a threat, the other player may be better off if he can somehow avoid communication so that the threat cannot be explicitly delivered.

In examining the effects of information and communication, it is important to distinguish joint gains from gains by one individual at the expense of another. If players work with equal skill and determination in trying to take advantage of each other, then their maneuvers may simply neutralize each other, leaving only the wasted effort and missed agreements as losses for each. With this warning in mind, let us now begin by examining some benefits of ignorance.

Knowledge is power, but can ignorance really be bliss? The fruits of knowledge are obvious in such forms as engineering, espionage, and market research, but where are the fruits of ignorance? One answer is that if you are a difference-maximizer, you may be happier not realizing that your neighbor is wealthy. On a larger scale, ignorance precludes a "revolution of rising expectations." Poor people who see only other poor people may not be dissatisfied with their lot until they find out how the "other half" lives. Unkept promises, another form of communication, can heighten the effect.

Ignorance affects utility in the foregoing examples, but it can also have strategic consequences. In some states and countries cars must yield to pedestrians in a crosswalk. They must but they *may* not, and if they do not it is the pedestrian who pays—a crumpled fender is negligible in comparison with a crumpled body. Nevertheless, in the "Crosswalk" game Pedestrian may win if she structures the information properly. Suppose that Pedestrian strides confidently into the crosswalk, chin up, nose straight ahead. Driver believes that Pedestrian does not see him, so he stops, to avoid an accident. This example involves *belief* about information: Driver believes Pedestrian is ignorant. In fact, the information interaction may be even more complex. Suppose Pedestrian is cleverly peeking out of the corner of her eye, just to make sure everything is working. Then she has information about Driver, specifically whether he *believes* Pedestrian does not know he is coming (as evidenced by his slowing down). Finally, if Driver is the Six Million Dollar Man and can see clearly from afar with his bionic eye, then he knows she knows whether he apparently believes that she apparently does not know that he is there (!) and so he can call her bluff by not slowing down. This game is not recommended for human beings, but shows that information and beliefs *about* information and beliefs can be interrelated in complex ways, even in ordinary situations.

Greed due to ignorance may be forgiven. Section 9.4, for example, presents the "Wholesaler-Retailer" game, in which Wholesaler unwittingly sets such a high price on some goods that Retailer can make only a small profit on resale. Retailer must then decide whether to buy or not. If Retailer can still make some profit, even buying at such a high price, then profit-maximization dictates making the purchase. Retailer might wish to punish Wholesaler, especially if he thought Wholesaler were deliberately taking advantage of him. Yet if he knows that Wholesaler acted from ignorance, not knowing Retailer's payoffs, then greed is not really an issue and the purchase will presumably be made. Both for Wholesaler here and for Pedestrian in the preceding example, ignorance, to be effective, must be known to the other player.

A player may even be ignorant of his own payoffs in the sense that although he knows what he likes he does not know the actual effects of his actions. This would occur, for example, if a person cared about not causing pollution but could not find out which new cars produced how much of which pollutants. He might also be happier not knowing that the cheapest car is not the cleanest.

In the remainder of this section, and in the next chapter, payoffs will be expressed in graph form. Misperception and misrepresentation of payoffs as well as threats and threatening acts then become *shifts* and *cuts* in these graphs. This visual approach will clarify the relationship among individual advantage, self-protection, and negotiated joint benefit. Communication will always be possible, though players may choose not to communicate. We shall begin with a simple example before moving to weightier affairs.

Game 5.4 Negotiating a Sale

Agatha advertises her petrified lotus leaf and only Bertram responds. The leaf is worth $100 to Agatha; she would not sell it for less. To Bertram the leaf is worth $200; he will not pay more.

Clearly Agatha and Bertram can do business together. Any transaction at a price between $100 and $200 is of benefit to both of them. If each of their current conditions is arbitrarily assigned a value of 0, then, for example, a sale at $130 is a gain of $30 for Agatha ($130 minus the leaf's value of $100 to her) and a gain of $70 to Bertram (the leaf's value of $200 to him minus the loss of $130). Any transaction will yield a combined gain of $100 (in this example, $30 + $70); that is, the value of the leaf increases by that much by shifting to the possession of someone who values it that much more. The situation is represented in Graph 5.1, where the slanted line, called the *negotiation set,* contains all pairs of payoffs that can result from a sale. Notice that the verbal description of this game translates directly into a graph, bypassing the matrix form. (It is left to the reader to ponder the difficulties of expressing the game as a matrix.)

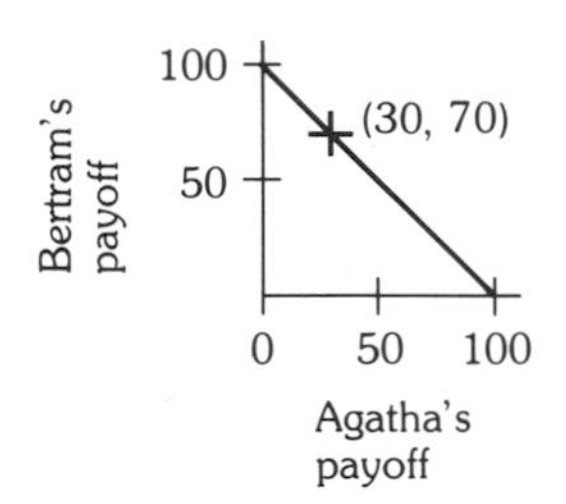

Graph 5.1 "Negotiating a Sale." The diagonal line, the negotiation set, contains all pairs of possible payoffs that can result from a sale. The point (30, 70) shows the payoffs for a sale at $130.

Payoffs have been expressed here in dollars rather than utility. As long as every dollar represents the same increment in utility, money is equivalent to utility with respect to an interval scale. This will be true provided that neither player can get noticeably more or less wealthy from a transaction.

An arbitrator might consider it "fair" to split the gains 50-50 in such a situation, setting the price at $150. Agatha and Bertram (acting on their own) might reveal their evaluations of the leaf and simply agree to such a deal, reasoning to themselves or to each other that no other single price is particularly likely to be agreed upon. There might even be a tradition in society calling for an even split in such circumstances. A tradition that called for uneven splits would have to include a way to decide who gets the larger share. In the theoretical treatment below, players are assumed to be equals, differing only by arbitrary names.

Uneven splits do seem plausible in real-life situations, however. One person may belong to some social class that is entitled to or expected to get the larger share in negotiable situations. Or it may simply be the case that one player is tougher, cleverer, or generally more skillful at negotiating than the other. Such skill might include the technique of misrepresenting one's utilities. For our graphs, matrices, and trees we typically assume that players know their own and each other's payoffs (though in matrices one player does not know the other's choice). In reality, a bargainer can easily lie during bargaining. This has been done for centuries in remote village markets the world over and in international parleys. In our example Bertram may do well to insist that he could not possibly afford to pay more than $170, or even $140. He can even make a legally binding promise to a third party that this is the case. By burning his bridges behind him in this way, he may further strengthen his bargaining position (for considerations of this sort, see Schelling, 1960, especially p. 24). Of course, if Agatha is equally aggressive, the deal may fall through.

Let us see what some of these stratagems look like in graph form. First of all, suppose that Bertram succeeds in misrepresenting to Agatha what the item is worth to him, so that she thinks he values it at $140. Agatha, on the other hand, reveals her true utility scale. The resulting payoff possibilities then appear to be those shown in Graph 5.2. Notice that what Bertram has done is to cause a down-

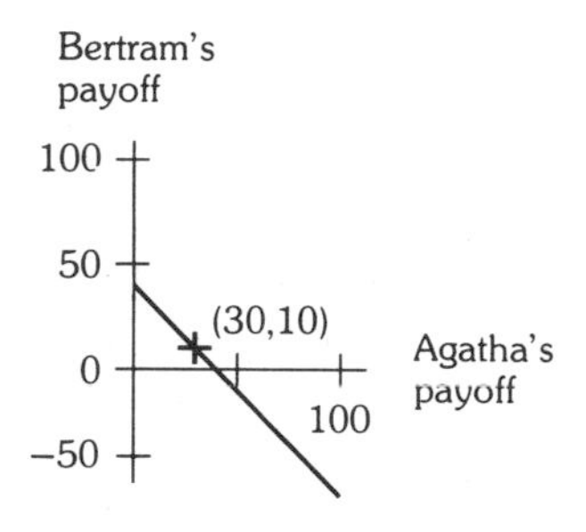

Graph 5.2 Payoffs for "Negotiating a Sale" when one party misrepresents his utility scale.

ward shift of the negotiation set (the line containing the various outcomes of the transaction). A transaction at $130, for example, now yields apparent payoffs of $30 to Agatha, as before, but only $10 ($140 − $130) to Bertram. The resulting point (30, 10) is shown on the graph, as is the no-transaction point, which remains at (0, 0). The apparent-payoffs point (30, 10) may look attractive to Agatha, who thinks it gives her better than half the available gain. In reality, this point is (30, 70), presumably satisfactory to Bertram.

To see what shifting sands bargaining is based on, suppose for a moment that Bertram's *true* bargaining position is $140, that the leaf really is worth only that much to him. How can Agatha know this? This consideration suggests that even a truthteller cannot expect to be believed. Next, suppose that Bertram does *not* misrepresent his payoffs but instead makes a legally binding promise to some third party, say Calvin, committing himself not to pay more than $140 for the leaf. Notice that he does not promise that he will necessarily succeed in getting the leaf, only that if he does get it the price will be $140 or under. Suppose that the agreed penalty that Bertram must pay to Calvin should he break the agreement is $75. The resulting payoff possibilities, assuming no misrepresentation by either player,

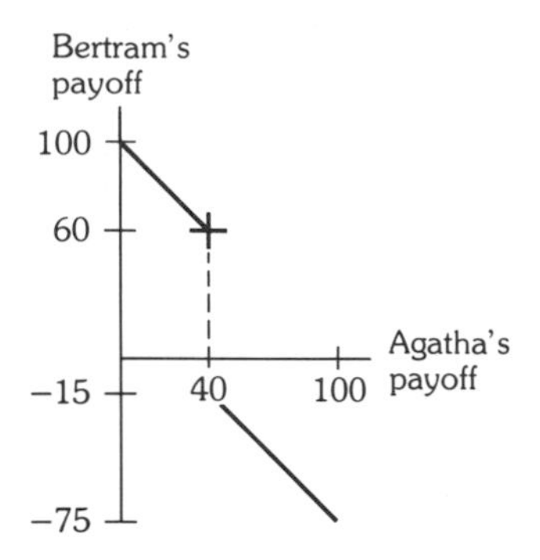

Graph 5.3 Payoffs for "Negotiating a Sale" when one player makes a binding commitment to a third party.

are shown in Graph 5.3. A transaction at $130 yields the same payoffs as in Graph 5.1 (30, 70), but now a high price will cause a truly negative result for Bertram. For instance, a transaction at $160 would yield a $60 gain in value to Agatha and a $40 gain in value to Bertram, the latter to be offset by a $75 payment to Calvin for a net loss to Bertram of $35. Thus, Bertram must convince Agatha that she had better settle for $140 since he truly, at this point, cannot profitably offer more. There is a small problem here, however. Although Bertram is really telling the truth this time, Agatha may or may not have ways of knowing that this is so. If she can verify that the deal with Calvin is genuine, and if she knows enough about Bertram's value system to be confident that the $200 is an accurate statement of what the item is worth to him, then her dollar-maximizing sale will be at $140. Whether or not she will go through with it (and possibly get a reputation as one who can be manipulated in this way) we cannot tell.

Now imagine that Agatha and Bertram are engaged not in buying and selling but in forming a business partnership. The purpose of the partnership is to make a business deal with a third party, Chloe. Agatha and Bertram's partnership will be assured a profit of exactly $100. Their payoff possibilities are again as shown in Graph 5.1. They can negotiate any split of the profits they both agree to, then both must sign the contract, which is then submitted to Chloe, the other party to the deal. The contract with their two signatures must reach Chloe by tomorrow morning to be valid. This partnership situation has the exactly the same joint-payoff graph as the previous game (Game 5.4). Our purpose in introducing it is that it will provide an appropriate context for the next stratagem.

In this new situation suppose that Agatha wanders into the kitchen to make coffee, and that while she is there Bertram writes in indelible ink on the contract an extra clause stipulating that he gets at least $60 out of the $100. He signs his name, leaves the document on the table, and drives off into the setting sun so that Agatha cannot possibly find him before the deadline. Now Agatha can either sign, thereby submitting to Bertram's manipulation, or she can let the deal fall through, giving up the $40 but perhaps keeping her dignity. The resulting payoffs are shown in Graph 5.4.

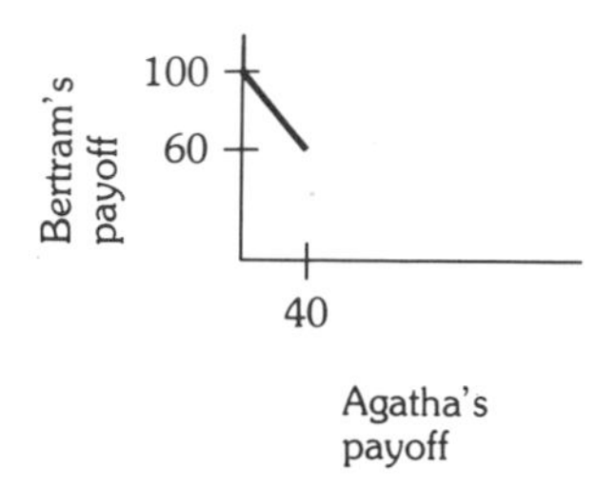

Graph 5.4 Payoffs for "Negotiating a Sale" when one party stipulates conditions and then breaks off communication.

In each of the three stratagems just described, represented by Graphs 5.2, 5.3, and 5.4, respectively, Bertram has arbitrarily been chosen as the perpetrator or engineer of the cleverness. If Agatha beats him to the punch, then *she* may benefit, and the cutting or shifting of the negotiation set must be adjusted accordingly, as shown in Graph 5.5.

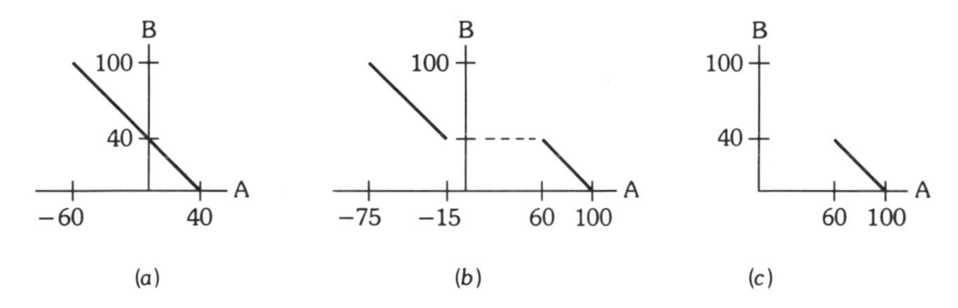

Graph 5.5 Payoffs for "Negotiating a Sale." Matrices (*a*), (*b*), and (*c*) show what happens when Agatha (instead of Bertram) engages in the stratagems of Graphs 5.2, 5.3, and 5.4, respectively.

But there is yet another possibility, namely the simultaneous use of one of these techniques by both players. For example, in the buyer-seller situation both players can misrepresent their payoffs or both can go off and make separate commitments to Calvin. In the partnership situation Agatha may have gone into the kitchen for the purpose of writing her own altered contract, and is slipping out the back door even as Bertram is leaving by the front.

What happens in each of these three cases? First, if both misrepresent their payoffs and these false payoffs are used to calculate a negotiation set, then all of its points will be unsatisfactory to at least one of the players, as shown in Graph 5.6a. Next, a pair of separate binding commitments would create Graph 5.6b, in which all parts of the negotiation set are worse for somebody than the starting point.

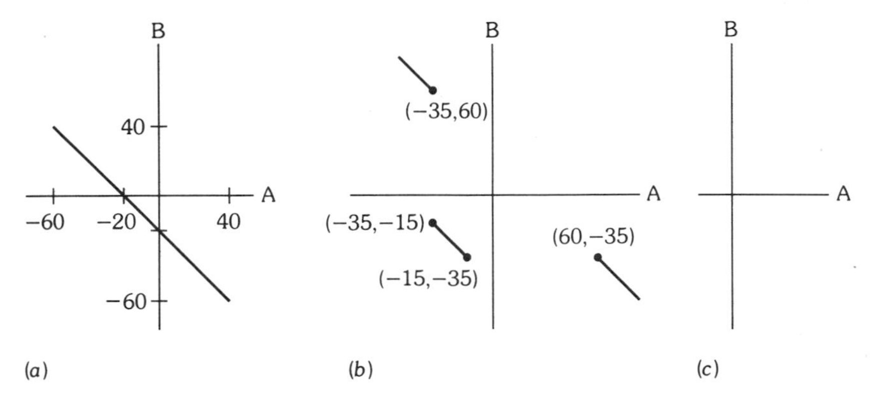

Graph 5.6 Payoffs for "Negotiating a Sale." Matrices (*a*), (*b*), and (*c*) show what happens when *both* players engage in the stratagems of Graphs 5.2, 5.3, and 5.4, respectively.

Finally, if both alter the contract and depart, hoping the other will cosign, both again lose out since no contract will get signed and delivered. In fact, the two separate contracts are not even compatible, since one requires the price to be above $160 and the other requires the price to be under $140. In game-theory terms, each player has wiped out more than half the negotiation set, leaving only the empty Graph 5.6c.

All three of these stratagems have two properties: any stratagem can be used by one player to aid himself at the expense of the other, but if used by both players, then both lose an opportunity for gain. In other words, according to the first property, these stratagems can yield zero-sum gains. The possibility of joint loss, the second property, is the variable-sum aspect of the situation. Strategic thinkers who advise us to use such stratagems, whether in personal or international dealings, are necessarily focusing our attention on the first rather than the second property. In this sense, such strategists are aptly called "zero-sum thinkers."

Exercises

○ **1** **(a)** Suppose that in Game 4.2, "Noisy Neighbors," there is an intermediate alternative available to each player, namely, to be mildly noisy, thereby getting only half the pleasure but also causing only half the annoyance to the other player. Give the resulting 3 × 3 matrix. (This will be a game with gradations.)

(b) Suppose that in Game 4.2 each player were to use a (1/2, 1/2) mixed strategy. What would the resulting payoffs be? What cell in the matrix of part (a) are they equivalent to? Why should this be so?

○ **2** In Section 5.2.2 repetition was incorporated into a matrix. It is also possible to incorporate the notion of binding commitment into a matrix (see Section 5.3). A player can make a commitment, according to Schelling (1963, p. 150) by "visibly and irreversibly" reducing some of his own payoffs. The mere announcement of an intention can have some binding force of this sort if one's prestige or credibility is at stake.

(a) To take Schelling's numerical example, consider the matrix shown and suppose that the rules specify that B goes first. Give the appropriate tree.

	B	
A	2, 5	1, 0
	0, 1	5, 2

(b) What can B do to get a payoff of 5 in this tree, assuming A is "rational," and what does A then get?

(c) Suppose A commits himself to the bottom row by causing 5 units to be subtracted from his top-row payoffs. What will B do in the resulting game (with B again going first), assuming A is rational, and what will A then get?

(d) Give a tree representing the whole situation: player A choosing to commit himself to the bottom row or not, then B making a choice in the resulting game, then A making a choice. An example of a strategy for A in this new (large) game is: "Commit, and then use the bottom row whatever B does." There are eight possible strategies (complete game plans) for A and four for B.

○ **3** Suppose nations A and B are negotiating a treaty. Let the current situation be assigned a utility scale value of zero for each of them.

(a) Along comes a mediator and finds an issue that A cares about much more than B. In fact, a concession by B loses him 1 unit and gives A a gain of 3. Show this shift on joint-payoff axes.

(b) The mediator also finds another issue for which a concession by A loses him 1 unit and yields a gain of 3 to B. Starting from the point achieved in part (a), show where this new concession gets the joint payoffs to.

○ **4** Present a non-zero-sum situation based on a newspaper or magazine article (which should be submitted), in which an important feature is one of the following: timing, gradations, repeated play. Your presentation should consist of:

(a) A fairly short verbal description of the particular aspect of the situation that you are focusing on, including some mention of either timing or gradations or repeated play.

(b) A tree, matrix, or other formal means of displaying the players, their alternatives, the sequence of moves, and the possible outcomes (states of the world).

(c) Some indication of the basis of preference for each player, and a preference order or an assignment of utilities.

(d) A discussion of the strategic considerations facing each player, with particular reference to timing or gradations or repeated play.

○ **5** A circumstance of play that is related to the notion of binding commitment is the degree of difficulty involved in reversing a decision after it has been made. Some decisions can be changed more easily than others because of organizational, political, or technological reasons.

(a) Give an example where lengthy deliberations preceded a decision because, once made, the decision would be hard to reverse.

(b) Give an example of a course of action that was pursued too long because of the costs of changing it.

(c) Give an example of a decision that is easy to change.

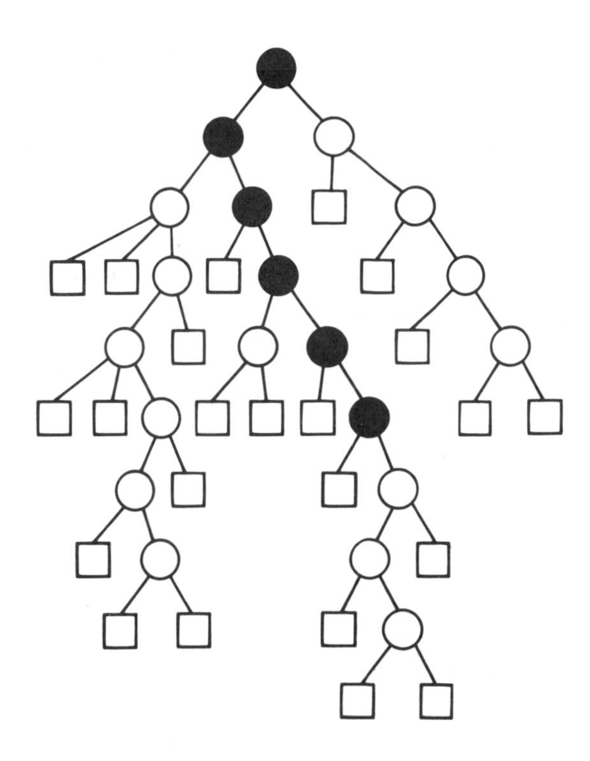

6 Approaches to Resolution of Two-Person Conflict

Can mixed-motive games like those in Chapter 4 be "solved" the way we solved zero-sum games in Chapter 3? The conflict between individual rationality and group rationality discussed in Section 4.2 suggests that the answer to this question must be no. In a sense the trouble may be traced to our definition of the word "solution." Solving a two-person, zero-sum game means finding the best payoff for each individual and how to get it.

The problem needs to be reformulated for variable-sum games, since what you can get on your own may not be as much as what you can get by coordinating with the other player. Indeed, the other player may be quite willing to make a deal, not through altruism (which would be incorporated into the payoffs), but because there is partial overlap of interests. In Section 6.1, a "solution" is redefined as a statement of how that partial overlap can be fairly utilized.

Some conflicts arise because one party finds it convenient to damage another. A factory that pollutes a river used by its neighbors is an example. If the damage is great and the remedy cheap, the neighbors may be willing to pay the cost of cleanup, thereby improving everyone's environment. In a sense, then, there is a partial overlap of interests. However, one may feel that factories ought to do their own cleanup, and therefore that governments should tax or regulate pollution. This issue and refinements of it will be treated in Section 6.2, using concepts discussed in Section 6.1.

6.1 Negotiated Solutions and Fairness

The solutions to zero-sum games in Chapter 3 were based on individualistic rationality principles, with each person pursuing his own benefit. In the case of "Prisoner's Dilemma" (Section 4.2.1) the result of using those very same principles was a deficient outcome, unsatisfactory to both parties. The weaknesses of our individualistic rationality principles were also evident in the discussion of "Chicken" in Section 4.2.3, where a whole series of considerations of individual motivations proved inconclusive.

Since neither player acting alone can solve these problems, the two players must somehow reach a coordinated pair of strategies. It is conceivable that coordination could be achieved tacitly, but common sense and several experiments indicate that the opportunity to communicate will be a big help. In addition, each player must somehow become confident that the other will carry out his part of the bargain. Such confidence can be based on enforcement or at least surveillance, where possible, or on trust. Trust may be based on someone's past performance if there have been repeated plays of the game (Section 5.2).

We will assume that, one way or another, agreements once made will be carried out. This leaves the following problem. Suppose the players in a two-person, variable-sum game are allowed to negotiate. What should they end up agreeing to? By use of the word "should" we seem to be making a moral pronouncement. Perhaps a more neutral, scientific question would be, What are they likely to end up agreeing to? If instead of having the two players themselves try to work out an agreement, we turn the decision over to a neutral party, we have converted a negotiation problem into an arbitration problem. In that case we may ask how the arbitrator is to proceed. Are there some fixed principles of fairness that she can turn to or should she impose a settlement according to what the players themselves probably would have (perhaps less efficiently) finally come up with?

Whether we take the view of a participant, arbitrator, or social scientist, the basic question remains, How do relationships among payoffs affect what should or will happen when a variable-sum game is negotiated or arbitrated? The answer to this question is what game theory dignifies with the title "solution" in the variable-sum case.

6.1.1 Payoff Graphs for Two-Person Games

To pave the way for an understanding of the concept of solution presented in the next subsection, we shall expand upon the graphs of Section 5.3. Recall that these graphs show the possible joint payoffs of the two players. For vividness let us use a current event:

● **Game 6.1 Rhodesia, Fall 1976**

The U.S. casts certain votes in the U.N. Security Council that are beneficial to the U.S. but more so to South Africa. The latter can aid the U.S. at its own expense by doing the United States' bidding in Rhodesia. (Specifically, it can threaten Rhodesia with economic sanctions or military supply cutoff, demanding the white regime to move toward black rule. This serves U.S. objectives but is not something South Africa wants to do.)

This game can be put into the form of a matrix with separable payoffs (see Section 4.2). Suppose that each player's payoffs fit an interval scale. Suppose that for the U.S. to maintain its voting policy is worth 2 to itself and 3 to South Africa, in comparison to zero for each if the U.S. votes differently. On the other hand, for South Africa to threaten Rhodesia is worth, let us say, −2 to South Africa and +2 to the U.S. (Since the outcome of such threats is uncertain at the outset, these numbers can be regarded as expected payoffs, which take into account the probability and desirability of various possible outcomes.) When these separate parts of the payoffs are added together, the result is Matrix 6.1. This game exemplifies two important concepts introduced in Chapter 4: it is separable and it has a threat-vulnerable equilibrium.

Let us now translate the matrix into a graph, in particular into what is called the *joint-payoff graph*. Each cell in the matrix becomes

		South Africa	
		2, −2	**0, 0**
U.S.	**2, 3**	4, 1	2, 3
	0, 0	2, −2	0, 0

Matrix 6.1 "Rhodesia, Fall 1976."

a point on the graph. For example, the payoffs (4, 1) in the upper-left cell map into a point with horizontal coordinate 4 and vertical coordinate 1 (point S). The players acting together can clearly achieve any of the four points P, Q, R, and S, simply by agreeing to choose the appropriate row and column, e.g., upper row and left column for (4, 1), which is S. They have other options as well. If the U.S. uses

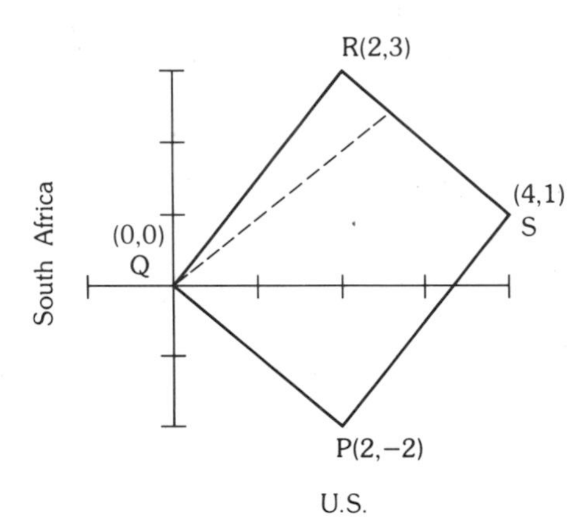

Graph 6.1 Payoff graph of Matrix 6.1.

the top row while South Africa uses a (1/2, 1/2) mixture of columns, then the U.S. gets an expected payoff of 1/2(4) + 1/2(2) = 3 while a similar computation for South Africa yields 2. This payoff pair (3, 2) is represented by a point midway between R and S. As another example, if both players use a (1/2, 1/2) mixed strategy, then each of the cells has probability 1/4 and the expected payoffs are (2, 1/2), corresponding to the center of the parallelogram PQRS. Every point inside or on the edge of that parallelogram corresponds to some pair of strategies, mixed or pure. For example, strategy pairs have already been found for (2, 1/2) on the inside and (3, 2) on the edge.

Now consider the two-person version of "Let's All Have Fun Doing It My Way" mentioned in Section 1.2 and called "Battle of the Sexes" by Luce and Raiffa (1957). The payoffs assigned by those

	B	
A	2, 1	0, 0
	0, 0	1, 2

Matrix 6.2 "Let's All Have Fun Doing It My Way."

authors are shown in Matrix 6.2. The resulting figure in Graph 6.2 is a triangle, because two cells in the matrix are identical and so correspond to the same point. Unlike the preceding example, this triangle has points that do not correspond to any pair of strategies chosen independently by the two players. An example is the point (1 1/2, 1 1/2), which lies on the line UV. To see that no pair of strategies

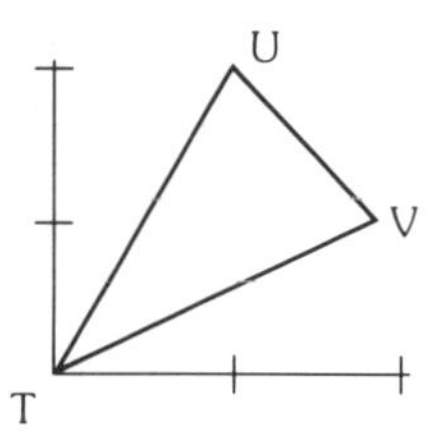

Graph 6.2 Translation of Matrix 6.2.

corresponds to this point, first note that unless at least one player uses a mixed (not pure) strategy the result will be a vertex of the triangle. However, use of a mixed strategy by either player creates a non-zero probability for one of the (0, 0) cells; therefore total payoff is sometimes 0 (= 0 + 0) and sometimes 3 (= 2 + 1 = 1 + 2), hence on average *less* than 3. But at (1 1/2, 1 1/2) the total payoff equals 3. This completes the proof that no pair of strategies yields the payoff pair (1 1/2, 1 1/2).

There is, of course, a way for players to achieve expected payoffs of 1 1/2 each. They can agree to coordinate their strategies, jointly arranging the upper-left or lower-right cell, each with probability 1/2. That is, they will flip a coin and whoever wins the toss will pick the cell, while the other agrees to go along with this result. In a real-life situation this is a fairly simple and reasonable way to resolve a conflict. More generally, players can jointly obtain any point in the triangle of this example or any point in the payoff *polygon* of other examples. Larger matrices may contain more than four distinct payoff pairs, each determining a point on the payoff graph. The payoff polygon or *convex hull* can be drawn by connecting all the points and shading in all of the enclosed area (some of the points and lines may end up inside the convex hull).

Before leaving the game of Matrix 6.2, it will be of interest to see what choices would be predicted by a "reinforcement" theory of behavior. Suppose that players successively make small adjustments in their individual strategy mixtures, each seeking increased payoffs or "reward." In no way can this lead to payoffs of 1 1/2 each. In fact, if each player starts with a (1/2, 1/2) mixture, player A will find a small shift toward the top row (say, to .51, .49) advantageous, while B will similarly find a small shift toward the right column advantageous. The joint effect will be *dis*advantageous to both! In "Prisoner's Dilemma" this sort of dynamic model typically leads to a deficient outcome (see Rapoport, 1965, Chapter 10).

The moral of the story is that *coordinated* strategy choices are essential to reaching certain regions of a payoff graph; in "Battle of the Sexes" this includes the point that is a reasonable candidate for the title of "most equitable," namely (1 1/2, 1 1/2). It is also important to remember that a graph does not give a complete picture of a matrix game. In going from matrix to graph, we lose information as to which cells were in the same row or column. By way of comparison, Matrix 6.3 has the same triangular graph as Matrix 6.2, but now the

	B	
A	1, 2	2, 1
	0, 0	0, 0

Matrix 6.3 A matrix with the same payoff graph as Matrix 6.2.

upper-right edge, including point (1 1/2, 1 1/2) *can* be reached without coordination. Despite the fact that the two games correspond to the same graph, they are vastly different strategically, as you can verify by searching each for dominant strategies, equilibrium outcomes, and threat-vulnerability.

6.1.2 Bargaining Principles and the Solution

It is time now to return to our search for a "solution" to the problem of what should or will happen under negotiation or arbitration. We shall first describe certain "bargaining principles" and then show how these lead to a solution. This approach was first proposed and worked out by Nash (1950).

First, let us return to Graph 6.1 for "Rhodesia, Fall 1976." Notice that any point *off* the line RS is deficient since *both* players do better by moving to some point on RS. Take, for example, the point (2 1/2, 1 1/2). Only South Africa would be interested in moving to R from this point, while only the U.S. would increase its payoff by moving to S. Nevertheless, there *are* points on RS that are better for both players, e.g., the midpoint of RS, (3, 2), which can be reached by explicit or implicit agreement.

In general, any movement *upward and to the right* on a joint-payoff graph represents an improvement for both players, so any point from which such movement is possible is, by definition, deficient (see Section 4.2.3). Equivalently, any point not on the upper-right border of the payoff polygon will be deficient. Thus, given the opportunity to negotiate, we might expect or at least hope that players would reach this upper-right border. These nondeficient (or Pareto-optimal) points, taken together, constitute the *negotiation set*.

If we grant that bargainers will or should end up on the negotiation set, the next question is *where* on that set, that is, at precisely which point? The U.S. prefers point S, while South Africa prefers point R. Any movement along the line improves one player's outcome at the expense of the other.

It is useful to turn away momentarily from the end of negotiations to the beginning. Negotiation is a process; where does it start? There are several possible rationales for where the starting point might be. One possibility is the point that corresponds to the equilibrium in a game that has exactly one such point. The equilibrium in Matrix 6.1 has a strong claim to our attention since it corresponds to the intersection of dominant strategies.

Another approach to finding a starting point, or *status quo point* as it is usually called, is to imagine that a player begins by noticing how much he can guarantee himself. That is, a player might look only at his own payoff and see what his maximin mixed strategy will guarantee him, just as if he were in a zero-sum game (see Chapter 3). This amount is called a player's *security level*. The point on the graph whose coordinates are the two players' security levels is called the Shapley status quo point. In "Battle of the Sexes" in Matrix 6.2 each player has a mixture that guarantees a payoff of 2/3. In "Rhodesia, Fall 1976" in Matrix 6.1 each player's maximin is a pure strategy since each has a dominant strategy. If the U.S. uses the top row it is assured at least 2. South Africa can guarantee only 0 by using the right column (though it may hope to do better than this). This game therefore has (2, 0) as its Shapley status quo point.

Altogether, several ways of determining status quo points have been proposed, and we shall look at one more a little later. By mentioning more than one, both of them superficially plausible, we have already made it clear that finding a "solution" concept for mixed-motive games is a difficult problem, one about which "rational" people may hold differing opinions.

Even if a status quo point can be settled on, the problem remains of specifying how the bargaining procedure should, might.

or will proceed from that point to the negotiation set. To shed light on this aspect of the problem, let us look in again on Agatha and Bertram and their selling-buying game. Here, we shall ignore the possibility of the tricky maneuvers discussed previously (Section 5.3). That is, we shall assume that all payoffs are known and that there is no basis for the analyst to think that either player has a bargaining advantage. Such circumstances are called symmetric and are assumed to yield equal payoffs to the two players. We shall now state this assumption along with the principle of nondeficiency (Pareto-optimality) mentioned earlier.

◀ **Bargaining Principle 1**
The solution lies on the negotiation set, i.e., the solution is not deficient.

◀ **Bargaining Principle 2**
In a symmetric situation players get equal payoffs.

Agatha and Bertram are in effect faced with splitting $100. Notice in Graph 5.1 that the symmetric negotiation set is at a 45° angle to the horizontal axis (and also to the vertical). Starting from the symmetric status quo point, one moves upward and rightward at 45°, ultimately reaching the even-split point (50, 50) on the negotiation set. Bargaining Principles 1 and 2 are thus sufficient for determining a solution in this simple case.

Suppose that for some reason Agatha must not receive more than $180 on the deal, for a net gain of $80. For example, a law forbidding swindles might prescribe $180 as a price ceiling on petrified lotus leaves. The effect of such a condition is to remove part of Graph 5.1, yielding Graph 6.3a. The removed region represents bargaining outcomes that are not reasonable anyway, so it would

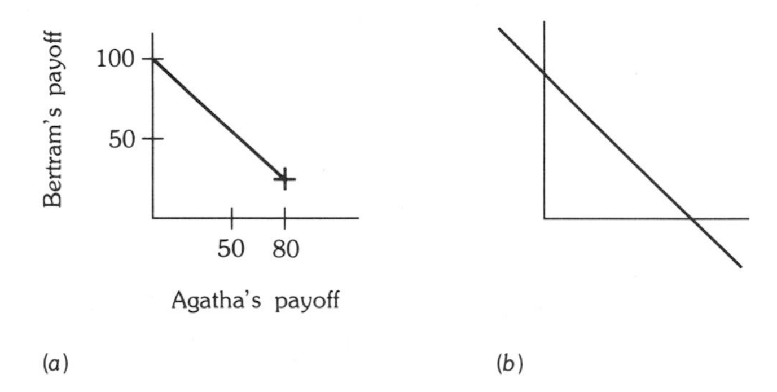

Graph 6.3 Graph (*a*) shows the effect of constraining the price of the lotus leaf to under $180. Graph (*b*) is formed from Graph (*a*) by extending the negotiating set.

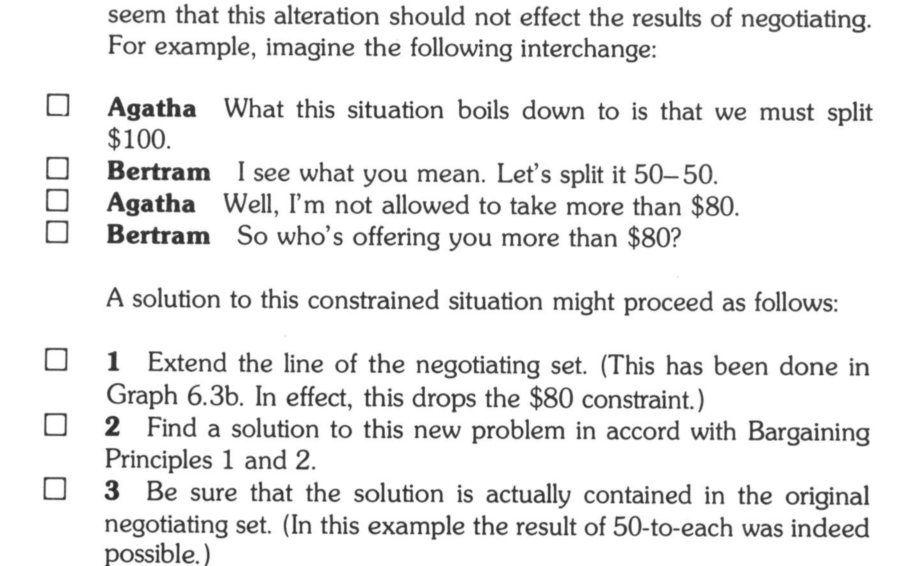

seem that this alteration should not effect the results of negotiating. For example, imagine the following interchange:

- ☐ **Agatha** What this situation boils down to is that we must split $100.
- ☐ **Bertram** I see what you mean. Let's split it 50–50.
- ☐ **Agatha** Well, I'm not allowed to take more than $80.
- ☐ **Bertram** So who's offering you more than $80?

A solution to this constrained situation might proceed as follows:

- ☐ **1** Extend the line of the negotiating set. (This has been done in Graph 6.3b. In effect, this drops the $80 constraint.)
- ☐ **2** Find a solution to this new problem in accord with Bargaining Principles 1 and 2.
- ☐ **3** Be sure that the solution is actually contained in the original negotiating set. (In this example the result of 50-to-each was indeed possible.)

Now consider the more general case in which the negotiation set may be any straight line from upper left to lower right, and the status quo point may be anywhere to the lower left of that line. We shall assume that payoffs are no longer in dollars but are measured on an interval scale of utility. In general the two players' scales may not be comparable, so that 10 units for one player need not mean anything like what 10 units means to the other. We will feel free to multiply one player's scale without multiplying the other player's, basing these manipulations on:

◀ **Bargaining Principle 3**
Solutions are unaffected by interval scale changes.

An example will show how the three bargaining principles enable us to solve the more general class of bargaining situations. Suppose that two people, Dorinda and Emmon, are in the situation represented by Graph 6.4. The status quo point Q is (3, 4) and the negotiation set consists of the points on the line segment between the points M (1, 16) and N (6, 1). We shall not ask what method was used to get the status quo point, but for now simply accept it. This graph might be a direct translation of a real-life situation so that the points (1, 16) and (6, 1) are states of affairs in the world. Alternatively, these pairs of numbers may have been cells in a matrix. Either way, since the numbers are utilities on an interval scale, we can give a lottery interpretation to other points on the line MN (see Section 3.2.2). For example, (3, 10) is regarded by both players as equiva-

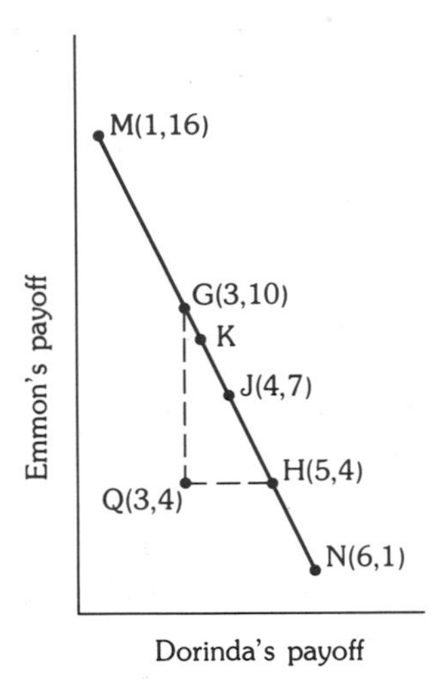

Graph 6.4 Finding a solution for the bargaining game.

lent to the lottery (3/5 M, 2/5 N), where M occurs with probability 3/5 and N with probability 2/5. This is true for Dorinda since $3/5(1) + 2/5(6) = 3$; it is also true for Emmon since $3/5(16) + 2/5(1) = 10$.

The status quo point may yield payoffs such that neither player need settle for less. In the case of Agatha and Bertram, either could call off the deal and be sure then of at least getting 0. In the case of Dorinda and Emmon, if the status quo point Q was determined by establishing the security levels of the players (the Shapley status quo), then it too represents payoffs that individuals can be assured of. If this is so then the status quo point (3, 4) means that Dorinda must get at least 3 (otherwise she will simply get 3 by acting independently). From the graph it follows that Emmon therefore cannot hope to get more than 10. The vertical dashed line from Q up to point G on the negotiation set shows how to find this "cutoff" point. Similarly, the horizontal dashed line yields another cutoff at H or (5, 4); this means that Emmon acting alone can insure 4 so Dorinda cannot reasonably hope for a bargain that gives her more than 5.

Serious negotiating thus is restricted to the region between G and H. Where in this "serious region" will or should Dorinda and Emmon end up? The midpoint of GH, namely point J, is an answer that is consistent with all three of our bargaining principles. In particular, it is unaffected by changes in the interval scale, so that Principle 3 is not violated. Principle 1 is not at issue since J is clearly on the negotiation set. Finally, Principle 2 concerns symmetry and the word "midway" certainly sounds symmetric. In fact, any other point, say

K, would violate Principle 2 in the sense that, with respect to points G and H,

☐ **1** K is more like Emmon's *more* preferred outcome (G);
☐ **2** K is more like Dorinda's *less* preferred outcome (G);
☐ **3** Therefore K favors Emmon.

The following summary of the geometric construction of the solution emphasizes the fairness of the method by showing that it can be used with either player.

☐ **Step 1** Pick a player, *either* player. Call her/him Z.
☐ **Step 2** Draw a line through the status quo point Q parallel to Z's payoff axis. This line hits the negotiation set somewhere. Call the point R.
☐ **Step 3** Find the midpoint of QR. Call it S.
☐ **Step 4** Draw a line through S parallel to the other player's payoff axis. This line hits the negotiation set somewhere. Call it "solution."

If the method is applied to Graph 6.4 using Dorinda as "Z," the results are as shown in Graph 6.5a; if Emmon is "Z" we get Graph 6.5b. The important point is that the solution is the same.

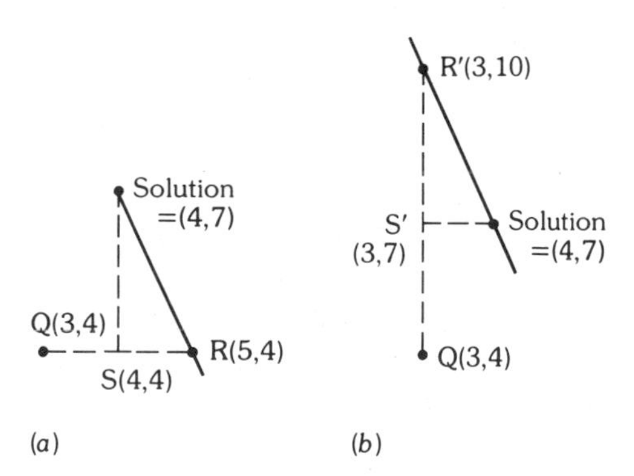

Graph 6.5 Two corresponding constructions that yield the same solution.

More complex negotiation sets can be handled within the conceptual framework already introduced. Thus, in Graph 6.6a the line AB can be extended as in Graph 6.3b, while the line BC is ignored, with the resulting solution shown in the graph. If, on the other hand, one ignored AB and tried to extend CB, the resulting point deter-

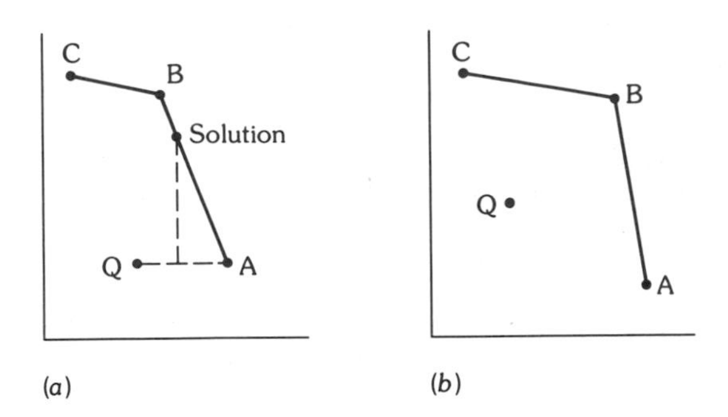

Graph 6.6 Solving the bargaining game when the negotiation set is not a straight line.

mined by the geometric method would lie to the upper right of the negotiation set, and hence could not possibly be reached.*

Fairness is not guaranteed by the symmetry condition (Principle 2). Consider two brothers: one a millionaire, the other a pauper. They jointly inherit $10,000, but in order to receive any of it they must agree on how to divide it. We will work in units of $1,000 for simplicity. To the millionaire, each $1,000 he may get will affect him equally. But the pauper's desperation for the first $1,000 exceeds his need for successive thousands. Let us imagine that on his utility scale the first $1,000 corresponds to an interval of 5, while subsequent thousands would be worth 4, 3, 2, 1, 1, 1, 1, 1, and 1. Then, starting from a status quo point of (0, 0), our solution technique given here awards $6,500 to the millionaire and only $3,500 to the pauper!

Is this result fair? Since the pauper has both greater needs and an equal legal right to the inheritance, he would seem entitled to at least half the money. To better understand this rich-get-richer distribution of the inheritance, note that each brother is happier with this result than he would be with flipping a coin for the whole $10,000. Looking at things a little differently, the game-theoretic solution indicates that the pauper's great need put him in a weak bargaining

---

* A more problematic situation is that depicted in Graph 6.6b. Here each line segment, when extended, calls for a solution beyond the negotiation set. We note, however, that among points on AB the point B is the most reasonable in the sense that it is closest to the point prescribed by the geometric method applied to the extension of AB. In the same sense, B is better than any point on BC. We conclude that it is appropriate to take B as the solution. In the case of a curved negotiation set, one may use a technique involving calculus. At each point on the curve, imagine a tangent line. Apply the geometric method. The solution will be that point at which the slope of the curve is the negative of the slope of the line connecting it to the status quo point.

position. Having gotten most of the *utility* he could hope for—5 + 4 + 3 + 1/2(2) = 13 out of a possible 20—the pauper is reluctant to jeopardize the bargain by demanding more.

6.1.3 Comparison of Status Quo Points

We shall now return to our original example, the threat game of "Rhodesia, Fall 1976" (Matrix 6.1). When we apply our methods of solution, starting from the Shapley status quo point, the results will seem a bit unfair to South Africa. Reexamination will show that the Shapley status quo point accords too much credibility to the U.S.'s threat. We will then introduce another method for finding a status quo point and show that it accords the threat a degree of credibility that depends on how much the U.S. has to lose by actually executing the threat.

First, suppose that the one equilibrium point of Matrix 6.1, point R, is taken as the status quo point. Since it is already on the negotiation set, R itself would then be the solution. Such a result in effect ignores the potential threat. The Shapley status quo point is (2, 0), the coordinates of the respective security levels of the two players, as noted earlier. Applying our method of solution to this point gives the outcome (3 1/2, 1 1/2), which is considerably closer to S than to R. It is, in a manner of speaking, "3/4 capitulation" to the threat, where R represents no capitulation and S is complete capitulation.

A comparable example with a more extreme outcome may make the point clearer. If the U.S.'s alternatives had (separated) effects of (3, 4) and (0, 0) rather than (2, 3) and (0, 0), then the payoffs would be those of Matrix 6.4, the Shapley status quo point

	2, −2	**0, 0**
3, 4	5, 2	3, 4
0, 0	2, −2	0, 0

Matrix 6.4 Variant of Matrix 6.1.

would be (3, 0), and the solution (5, 2). This result seems unlikely to occur in fact, since it is complete capitulation to a threat whose cost of execution is substantial (its cost is an interval of 3 for a player whose entire scale runs only from 0 to 5 in this situation).

An alternative approach to finding a status quo point, proposed by Nash, looks ahead to what the solution will be. It imagines that players jockey in advance for a status quo point that will give them as good a result as possible when that point is used to compute the solution by the method used above. The calculation of such a point turns out to be easy enough in the particular example at hand because the negotiation set happens to be inclined at a 45° angle to the axes. In other cases it may be necessary to change the interval scale for one of the players in order to obtain a negotiation set with a 45° inclination.

When the negotiation set is inclined at a 45° angle to the payoff axes, the solution can be found by moving from the status quo point upward and rightward at a 45° angle until the negotiation set is reached. This method is equivalent to the four-step geometric construction given earlier. (This can be seen from elementary geometry; the demonstration is not included here.) Graph 6.7 shows five possi-

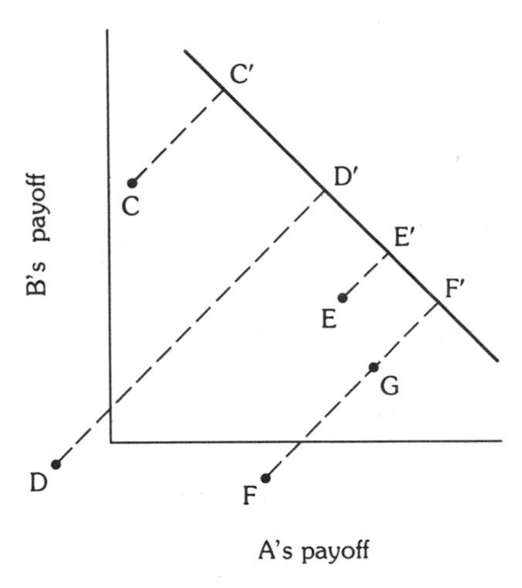

Graph 6.7 Various possible status quo points and the corresponding solutions.

ble status quo points. Points F and G will result in the same solution, namely F′. Each player is indifferent between these two as status quo points, even though each player would much prefer G to F if these were taken as *final* payoffs, from which no negotiation and no movement could occur. The worst status quo point for player A is C since it results in the solution C′ on the negotiation set, and C′ is to the left of the other possible solution points D′, E′, and F′. On the other hand, the solution point C′ is better than the other possible solution points for player B since it is higher.

These examples help show that if the players are jockeying for the selection of a status quo point their interests are necessarily diametrically opposed. Whatever change helps one will hurt the other. Change to the upper right, e.g., F to G, has no effect, nor does change to the lower left, G to F. Thus, a player has not jockeyed effectively if he merely moves the status quo point in such a way that he and the other are benefited equally. Rather, he must seek *relative* advantage, which may mean getting to a status quo point that is bad for himself but *very* bad for the other. For example, in Graph 6.7 point D is better than E for player B since D′ gives him a better final result than E′.

All of these considerations are captured by the *difference matrix,* found by subtracting one player's payoff from the other's in each cell. The difference matrix of Matrix 6.1 ("Rhodesia, Fall 1976") is shown in Matrix 6.5, where, for example, 4 in the lower-left cell is the

3	−1
4	0

Matrix 6.5 The difference matrix of Matrix 6.1.

difference between the payoffs (2, −2) in the lower-left cell of Matrix 6.1. An optimal strategy in the difference matrix will do just what is needed in jockeying for status quo, namely seek relative advantage. We can therefore treat the difference matrix (Matrix 6.5) as a zero-sum game. Whatever mixtures solve the (zero-sum) difference game are then used in the original variable-sum game to determine the Nash status quo point.

A strong cautionary note is in order here. This method of forming a difference matrix must be used *only* when the negotiation set is inclined at 45° to the payoff axes. Otherwise, one player's payoffs must be subjected to an interval scale change to achieve a 45° inclination. The resulting payoffs must be used in forming the difference matrix.

This particular difference matrix has a dominant strategy for each player: bottom row and right column. Using these choices in the original Matrix 6.1 gives us the Nash status quo point of (0, 0), the lower-right cell. The resulting solution of (2 1/2, 2 1/2) is relatively close to R; it represents a "1/4 capitulation." Thus, in this example the Nash method of finding the status quo point in effect accords less credibility to the threat than does Shapley's method.*

---

* One must be very cautious about interpreting the numbers in a difference matrix. They are calculated from the interval scales of utility of two different people, and each scale can be separately transformed by multiplication and/or addition. Nevertheless, some people like to think of this method as breaking up the game into two parts: a competitive or zero-sum aspect, used for finding the status quo point, and the purely cooperative movement from the status quo point to the solution on the negotiation set.

For the sake of comparison, consider what would have happened if the (2, 3) option available to the U.S. had been, say, (1, 3) or (4, 5). Would either of these make the threat more or less credible in terms of how much the threatener would damage his own interest by actually executing the threat? And what results do the Shapley and Nash status quo points prescribe? Answers to such questions (left for the reader to calculate) will yield insight into the relative merits of the two methods.

The steps involved in finding the Nash status quo point are the following:

☐ **Step 1** Check that the negotiation set is inclined at 45° to the payoff axes. (If not, make it so by changing the interval scale for one of the players. Use the resulting matrix in Step 2, just below.)
☐ **Step 2** Form the difference matrix.
☐ **Step 3** Determine the maximin strategies in the difference matrix.
☐ **Step 4** Apply the strategies (found in Step 3) to the original game to get the Nash status quo point.

Rapoport (1960, p. 112) presents the intriguing example of Matrix 6.6, in which the same numerical values are available to the

	B	
A	2, 1	−1, −2
	−2, −1	1, 2

Matrix 6.6 Player A's threat mixture and security mixture are the same, but B's are not.

two players but in subtly different configurations. Each can guarantee himself 0 by an appropriate mixture, and each can keep the other to 0 by some mixture. But for player A these two things can be done simultaneously, while B can keep A to zero only by abandoning his own security. It may help to set out two definitions:

◀ **Security Mixture**
A player's maximin with regard to *his own* payoffs (completely ignoring the other player's payoffs).

◀ **Threat Mixture**
A player's maximin with regard to the negatives of the *other player's* payoffs (ignoring his own).

In these terms A's security and his threat are both the (1/2, 1/2) mixture. For B, however, security requires (2/3, 1/3) while his threat mixture is (1/3, 2/3). The Shapley status quo point of (0, 0) ignores this distinction, whereas in the Nash approach it is crucial. In the difference game formed from Matrix 6.6, A picks the top row and, no matter what B may do, the resulting status quo point will yield a solution (2, 1), giving A his best possible outcome.

In summary, we have seen that one can attempt to provide a solution intended to apply to all mixed-motive games. The idea is to set out principles or criteria of some sort—nondeficiency, common sense, fairness, credibility, realism, security—and to look for methods that meet some of these criteria. The mere existence of at least two alternative schemes, neither of which is particularly absurd, should show that the question of what constitutes a fair and realistic resolution can be a matter of opinion. Nevertheless, the formal analysis clarifies the issues by showing which criteria are compatible with others and what consequences they have in particular situations.

6.2 Solution by Government

In this section we shall analyze certain governmental solutions to see how they compare with the negotiated solutions of the preceding section. The conflicts that we shall consider are two-person situations in which one person damages another out of self-interest. Specifically, we shall discuss situations in which someone pollutes and someone else suffers the consequences. The government actions we shall look at include no action, banning pollution, placing a limit on it, and taxing it. After considering government action on its own, we shall take a look at how it interacts with the negotiation effects discussed in the preceding section.

Our examples will refer to pollution in general; the specific kind of pollution will not be of concern, nor will the nature of any technology available to control it. In fact, although we shall speak of pollution, we could speak in still more general terms of any person, business, or nation that, in the course of pursuing its own benefit, imposes uncompensated costs or damages on someone else. Thus, if I build a new extension on my house, blocking my neighbor's ocean view, the payoff structure will be no different than if I subject him to "noise pollution" by using my power lawn mower at 6 A.M.

Economists call such undesirable side effects "external diseconomies" or "negative externalities." These effects are "external" because they damage people who do not create them and thus are external to the decision-maker. ("Positive externalities" can also occur; for example, scientific research may benefit firms that did not do the research). From the viewpoint of the polluting party, if it is a profit-maximizing firm, the external costs do not enter into decision-making. From our viewpoint as citizens, however, if we are trying to design a better society, we may wish to propose laws that somehow cause *all* the effects of an act to be weighed in reaching a decision on it. Perhaps surprisingly, it turns out that negotiation alone can be sufficient to achieve "efficiency" in the economist's sense, i.e., maximum net benefit (Section 6.2.1). However, severe obstacles often keep negotiations from achieving either efficiency or a fair distribution of payoffs, so that there is, after all, a role for government (Section 6.2.2).

6.2.1 Laissez-Faire

● **Game 6.2 Polluting**

Polluter pollutes and Sufferer suffers the consequences. The (marginal) costs of pollution abatement are such that it is cheap to clean up a little; but it becomes harder and harder to get cleaner and cleaner, until the last little bit of pollutant is very expensive, if not impossible, to eradicate. The (marginal) damage to Sufferer is just the opposite. That last bit of pollutant at the clean end of the range causes little damage and might even go unnoticed. Successive additions of pollutant become more serious as one moves, perhaps, from aesthetic considerations, to inconvenience, to damaged health.

Polluter and Sufferer may be thought of as an upstream chemical plant and a downstream fishery, or an upwind paper mill and a downwind human being, or a noisy workshop next door to a doctor's consulting room. Polluter and Sufferer have interacting decisions and payoffs; at least Polluter's decisions obviously affect Sufferer's payoffs. But what decisions are available to Sufferer?

A common starting point in analyzing such situations is to assume that Sufferer can offer to bribe Polluter, that is, pay him to install abatement equipment. If the cost of the equipment is higher than the cost of the damage it can avert, then it will not be worth it to Sufferer to make such an offer and there will be nothing to negotiate. This seems reasonable in that even if Sufferer became the owner of the factory, so that Polluter and Sufferer were merged, he would not install pollution abatement equipment. If, however, the extent of damage is greater than the cost of avoiding it, common sense suggests that something should be done. Indeed, the device of the bribe will make action possible without government action, simply by a two-party negotiation. As in "Negotiating a Sale," the optimal action is clear: cleaning up or making the sale. Only the price (the size of the bribe) remains to be determined.

To explicate the relationship between this situation and a buyer-seller situation, we can make use of payoff graphs like those in

Section 6.1. To start with, imagine a payoff structure even simpler than that of Game 6.2, in which each successive unit of pollution has the same removal cost, *c*; in the terminology of economics, marginal cost is constant and equal to *c*. Let us also assume that a similar statement is true of damages, that each successive unit of pollution removed yields the same reduction in damage, *d*. Implicit in this statement is that successive reductions of damage can be compared to each other in value. One approach to comparing different kinds and levels of damage is to say that damage to a person equals the amount he would be willing to pay to avoid it.

The possible joint payoffs are shown in Graph 6.8, where U is an arbitrarily chosen point that reflects the payoffs when there is no

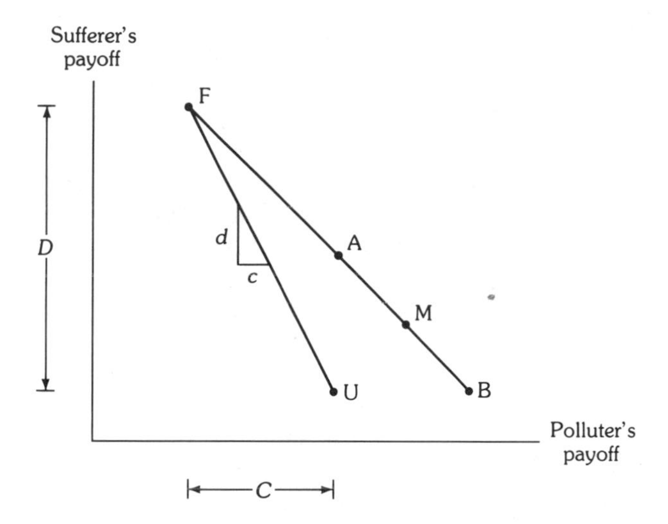

Graph 6.8 Joint-payoff graph for a simple pollution game with constant marginal cost and benefit. Point F represents full abatement and V no abatement. C = total cost of abatement (c = the cost of removing one unit of pollution); D = total reduction in damage (d = reduction in damage of one unit of pollution).

transaction, i.e., no abatement at all. Voluntary reduction by Polluter would result in movement from U toward F, or full abatement, along a line whose slope is $-d/c$. Both axes are measured in dollars. In exchange for abatement, Sufferer makes a payment. Each dollar paid by Sufferer is a dollar received by Polluter, so that we can represent potential payments as moves along the line FAMB, with slope -1, measuring payoffs in units of money.

In the absence of negotiation, and with no laws or taxation impeding his activities, Polluter may choose any point along the line UF and, unless he feels altruistic, will operate at U. If bribes are legal, Sufferer can offer to pay the total cost, *C*, of the cleanup operations.

This is a generous offer on the face of it. After all, it is Polluter who is doing the damage, and now someone else is offering to pay for the consequences.

In fact, however, a look at Graph 6.8 reveals that Sufferer must pay an amount greater than the cost of cleanup according to the negotiated solution of Section 6.1. If Sufferer simply pays for cleanup, the result will be point A, with Polluter neither better nor worse off than at the starting point U.* On the other hand, Sufferer's condition is improved by $D - C$ (D in relief from suffering less the payment C). By this reasoning, all the advantage of such a transaction goes to Sufferer, whereas according to the negotiated solution the improvement is to be evenly split. If Sufferer were to pay an amount D equal to the full value of being relieved of pollution damage, he would end up no better than he started, while Polluter would have a gain of $D - C$ (payment D less the cost C); this would be point B. The Nash solution is to split the benefits. This result is achieved if Sufferer pays $1/2(D + C)$ and is represented on the graph by M, the midpoint of AB.

It may seem unfair for Sufferer to pay for the whole cleanup and an extra amount as well. Our analysis need not be seen as denying this unfairness. What the analysis shows is that without government action Sufferer is in an exceedingly weak bargaining position. Therefore, one possible conclusion is that government must intervene.

The economic impacts of a wide variety of government policies can be analyzed using the concepts introduced in our pollution example. First, the sum of benefits, measured in dollars, is made as large as possible. To do this, all costs and benefits to all players are computed for a possible action, that is, the sum of all costs to both private parties and the government is subtracted from the sum of all benefits. Various possible actions can thus be compared. The cleanup action in our example involves a pure benefit D to Sufferer and a pure cost C to Polluter, for a net gain to society of $D - C$, which is better than 0, the result of inaction. This maximization may introduce certain inequities called *windfalls* and *wipeouts,* which are corrected (by a redistribution of money or other benefits) as part of the same deal.

Money plays a key role in all these considerations because it is freely transferable in any amount and allows one person's payoffs to be compared with another's. Although this simplifies things, it has a

* Recall that we have assumed there are no laws against polluting, so Polluter does not care how cleanly his factory operates.

serious consequence that is highly questionable. In effect, we award power proportionately to wealth: "one man, one vote" is replaced by "one dollar, one vote."

The solution in this case, just as in "Negotiating a Sale" (Section 5.3), relies on the assumption that payoffs are known. Here as there, a player may misrepresent his payoffs, threaten, and take action to render his threat credible. Recall that such measures tend to be advantageous if used by one party only; but they may be neutralized by the other and, if used too strenuously by both, may even undermine negotiations altogether.

Since the law in this example allows Polluter to do what he wants, one might imagine that Sufferer would have to open negotiations, thereby tacitly acknowledging that the pollution is bothering him. This may make it difficult for him to take up a sham bargaining position later on, claiming that it is not bothering him very much. On the other hand, Polluter may not have much chance to misrepresent payoffs since equipment costs are relatively easy to measure (still, Sufferer may not be knowledgeable about the technological details of Polluter's factory).

We now leave this simplified version of the situation and return to the somewhat more realistic Game 6.2, in which marginal costs and marginal damages are *not* constant. The situation can be represented by the economist's graph of marginal quantities, in this case marginal cost and marginal benefit of successive units of abatement. According to the description of Game 6.2, marginal costs rise as abatement proceeds, while marginal benefits fall (Graph 6.9). For

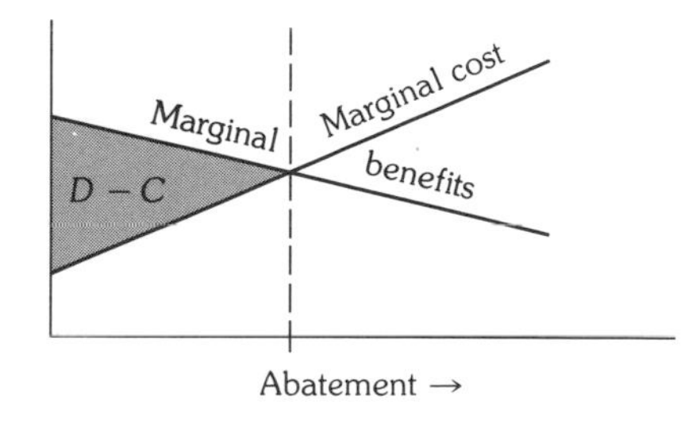

Graph 6.9 Net social benefit is maximized by abating to point I beyond which costs of the abatement project increase faster than benefits. (The point is the same whether abatement is regarded as a flow, measured in tons per day, or a stock, measured in tons for the project.)

the two-person society of Polluter and Sufferer, a net gain in combined payoff can be achieved by removing pollution as long as benefit exceeds cost for each unit removed, i.e., up to point I in Graph 6.9. Beyond this point additional abatement will cost more than it is worth, so that total social benefit is maximized at I. This much abatement yields an amount D of damage relief at a cost C for a net social gain of $D - C$ (the shaded area in the graph).

The joint-payoff graph (Graph 6.10) reflects the variability in marginal cost and marginal benefit (damage relief). Specifically,

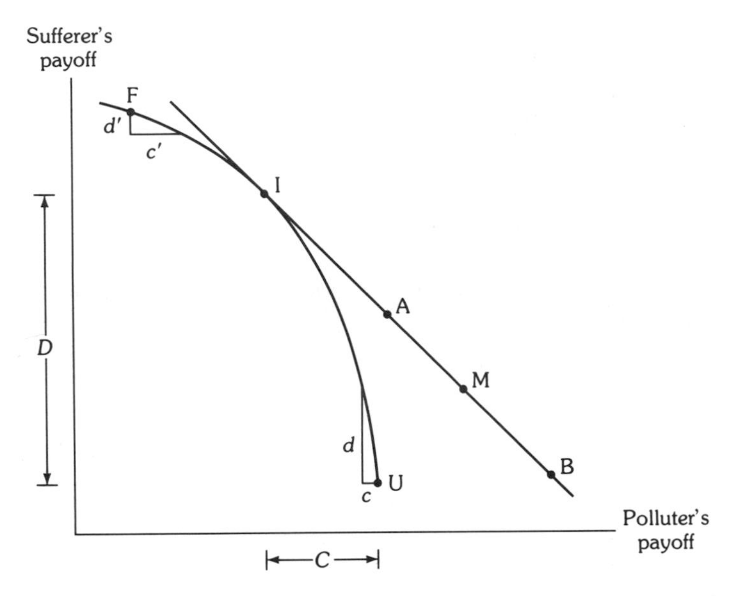

Graph 6.10 Joint-payoff graph for Game 6.2 in which marginal cost and marginal benefits behave as shown in Graph 6.9.

again letting U denote unabated pollution, and F the point of full abatement, Graph 6.10 presents a curve leading from U to F, where Graph 6.8 had a straight line. This curve reflects the assumptions of Game 6.2 that removal of the first bit of pollution can be done at low cost c to Polluter and yield substantial improvement d to Sufferer, but as things get very clean, further removal has relatively high cost c' and low benefit d'.

At point I on Graph 6.10 marginal cost equals marginal benefits, so the slope of the curve is -1. The dashed line tangent to the curve at this point also has a slope of -1. It contains all points that can be achieved by the two-stage process of having pollution abated to I, the point of maximized group benefit (with costs included as negative benefits), and then having Sufferer pay some amount of compensation to Polluter. This dashed line is the negotiation set, since any point on the curve UF or any point reachable from it by transfer of money is inferior for both players to some point on the dashed line. It remains to be determined just where on the negotiation set the players will reach agreement. The amount paid must be at least C if Polluter is to be interested (in the absence of legal restraint), and no more than D for Sufferer to be interested. The Nash solution is again a payment of $1/2(D + C)$, provided its assumptions are met.

6.2.2 The Right to Breathe

So far we have not brought the government into the picture, except to stipulate in our example that there is no ban, tax, or regulation on pollution. In effect, we have assumed a political condition in which there is a "right to pollute." Let us now turn the tables and assume a "right to breathe," or more generally, a right to be free from pollution. This right has a legal history dating at least from 1611, when "an English court granted an injunction and damage to a plaintiff whose air had been corrupted by the defendant's hog sty" (from Thompson's [1973] excellent chapter "Legal Approaches to Cost Internalization"). How is the game-theory solution affected by this change in legal assumptions?

As before, the parties will be free to negotiate whatever deals they can manage to agree to; for example, Polluter may wish to buy a limited amount of pollution rights from Sufferer. We assume that if no agreement is reached, no pollution at all is allowed, that penalties are severe, and that enforcement is perfect. To keep the negotiations flexible, imagine that Polluter is just building a new plant and is deciding how clean its operation will be.

All the potential joint payoffs are exactly the same as in Graph 6.10, since the physical situation is unchanged. In other words, the curve UF is unchanged and so is the negotiation set. The only difference is that now, in the absence of negotiation, Polluter must run a perfectly clean operation, so that the status quo point is now at F rather than U.

Once again the solution may be regarded as having two parts, one a change in factory operations causing a movement along the curve, this time from F to I, and the other a transfer of money. This time it is Polluter who must pay, and, as can be confirmed from the graph, the amount must be between V and W.

The above argument yields an intriguing result. The initial assumption of whether there is a right to pollute or a right to breathe affects only the payment of money and not the question of whether abatement equipment is installed, since in either case the factory operates at point I. This conclusion hinges on the assumption that there are no costs of bargaining. Bargaining is said to be costless if there is negligible time, money, or effort expended in reaching agreement.* Also, it is assumed the negotiators will reach the negotiation set, though not necessarily at the point denoting an even split of net benefit. This result seems to suggest that government action here, assigning rights of one sort or the other, has no effect other than to achieve an equitable distribution of money, something that a government might achieve more comprehensively by general taxation methods. However, there is a good deal more to the problem, as we shall see.

First of all, bargaining is not costless; it does take substantial time and effort, or if done by hired lawyers, money. Even if one is resigned to the need for such costs and goes ahead with the bargaining process, there is no guarantee that the negotiations will result in agreement.

In addition to the general problems of threat and misrepresentation that are present in general in bargaining, this particular situation has a threat potential of its own. That is, with no government regulation Polluter can decide to run his plant overtime. Even if this is unprofitable, it will nevertheless put him in a stronger negotiating position in terms of the theory in Section 6.1.2, provided he harms Sufferer more than himself. If Sufferer can be convinced that overtime is desirable to Polluter, he may even pay simply to restore normal working hours. In such a case Sufferer would have been better off if negotiation were illegal.

There is another drawback to letting Polluter and Sufferer work out whatever they can agree to, independent of governmental intervention. Our two-person game has not taken into account that in the real-world situation the role of Sufferer typically belongs not just to one individual but to several people in the neighborhood of the factory. In order for them to speak with a single voice at the negotiating table they will have to communicate with each other in some way, agree to contribute time or money in some fairly shared way, and establish just how great their marginal suffering is. If they do reach agreement among themselves, they will still face the problem that those who do not live up to their commitment may nevertheless benefit from the group's efforts. Thus, a group of sufferers is confronted with a payoff structure that strongly inhibits attempts at self-organization. In fact, the game the various sufferers play among themselves is none other than multi-person "Prisoner's Dilemma," which forms the basis of Chapter 7.

---

* Coase has written a seminal paper on the notion of costless bargaining. Critiques and extensions of this work have been summarized by Turvey. Both articles are reprinted in Dorfman and Dorfman's excellent reader (1972).

All of these bargaining difficulties could be avoided if the government simply established an emissions standard, prohibiting pollution above a level corresponding to I. The question is how the government can know what that level is. To find point I the government presumably would have to know the marginal cost and marginal damage functions in order to determine where they intersect. Such information may be hard to get, for you cannot simply ask someone how much he is suffering. Asked how much money it would be worth to stop being polluted, a person may give one answer if he thinks that you are going to compensate him by that amount and a very different answer if he thinks you are planning to charge him that much for pollution removal. Measuring costs may also present difficulties. In short, bringing in the government does not get around the problem of information and the possibility of misrepresenting it.

Nevertheless, the government might do well to set a standard at some moderate level, without worrying too much about whether the exact optimum has been found. A look at the curve in Graph 6.10 shows that it contains a wide range of points on either side of I that are all closer to the negotiation set (the tangent line) than F or U. Thus, if negotiation is considered unattractive for the reasons given above, a standard may be far superior to either a ban or no action.

In fact, government standards are implicit in certain longstanding laws on such things as nuisance, negligence, and abnormally dangerous activities. Moreover, court rulings in these matters have often hinged on just the kind of utility considerations we have been discussing. As early as 1904 a Tennessee court granted financial compensation to farmers whose crops were damaged by a sulphur mill (*Madison* v. *Ducktown Sulphur, Copper & Iron Company,* 113 Tenn. 331 [1904]; see Thompson [1973]). However, the court refused to shut down the mill, citing the relative financial values, and hence implicitly the *utilities,* of the adversaries. In the language of Graph 6.10, this ruling means that d was much smaller than c, so that movement away from U by closing the mill would be nearly horizontal. Financial compensation, which moves the parties along a 45° line through U, made more sense to the court.

Governmental standards have been criticized on the grounds that if Polluter is obliged to compensate for all the damage he does, then Sufferer, knowing he will be compensated, may deliberately expose himself to extra damage.* New "sufferers" might even move

* This discussion assumes that money taken as a penalty from Polluter will be given as a compensation to Sufferer. The government could keep this money (whether it is a fine or a tax, as below), but if it does so the maximum social benefit may not be achieved. For explanation, see Turvey, 1972.

into the neighborhood to take advantage of potential compensation. Society would thereby move away from maximization of total benefit. To take Coase's example, suppose that the role of Polluter is played by Rancher, and that of Sufferer by neighboring Farmer. Rancher's stray cows sometimes trample Farmer's fields. If the law requires that Farmer be compensated for any damage, he may do well to initiate production on otherwise unprofitable portions of his land. Such an action can be used to gain an advantageous starting position in negotiation. In this sense it is comparable to Polluter running his factory all night in the earlier example.

Many economists, perhaps most, would be displeased with the argument presented so far. They would be happy to see the government take action but would consider a standard or fixed limit on pollution to be a crude and rather ineffective device. Instead, they would call for a tax. A tax on pollution offers the polluter some flexibility: if he does not want to clean up he can pay the tax instead. A standard, in contrast, requires the same response from all polluters: reduce pollution to a certain level. The flexibility allowed by a tax is important since not all polluters will find it equally feasible to clean up. Those who can do so relatively cheaply will prefer to do so rather than pay the tax, whereas others whose technological situation makes cleanup expensive may opt for the tax. Each polluter's decision will be based on an economic analysis of his own situation, something that he knows far more about than a government agency could.

The problem of misrepresentation of information now disappears for the individual firm. Noxious Chemical Corporation does not have to try to convince the government how high their marginal cost for cleanup would be; if it really is high, they opt for the tax. However, the problem of misrepresentation reemerges at the next level of organization, since an entire industry certainly does have an incentive to claim that cleanup would be horribly expensive. A notable example is the decade-long controversy over automobile emissions and the feasibility of control devices.

The flexibility permitted by a tax exists also, in a sense, with standards. Just as a firm can opt for a tax, so it can opt to break the law and not conform to a standard. Thus, any analysis of standards must take into account the penalty for noncompliance and the likelihood that it will be invoked (see Section 4.1). However, even taking the possibility of noncompliance into account does not make standards as flexible as taxes, as the following comments show.

Taxes exert a continuous pressure toward improved performance. This cannot be said for standards—either you meet the standard or you do not. If you meet it, there is no incentive for further cleanup. If you do not meet the standard, but some very cheap device is invented that would get your pollution level almost down to the standard, there is no incentive to use it. A tax, in contrast, provides incentive at all levels. Since any decrease in pollution means a decrease in taxes, any cleanup device that is cheap enough will be used. "Cheap enough" means, roughly, that the cost of removal per unit of pollutant is cheaper than the tax (and cheaper than other devices). The life-span and operating costs of a device must be considered as well as the likelihood that a still better device may soon be available. Thus, the decision will not be simple, but the basic point remains that the tax puts every firm into the market for pollution equipment. The resulting effect on the manufacturers of such equipment cannot help but spur research on better devices.

Another type of government action, similar in some ways to taxation, is to offer every polluter a certain payment, say in the form of a tax rebate, for each unit of reduction in pollution. The prospect of getting a bigger payment, like that of paying less taxes, provides a continuous incentive to clean up. Payments, however, have some highly undesirable effects that taxes do not share. Payments would imply some "right to pollute," whereas taxes imply the "right to breathe." By starting the calculation from the original uncleaned-up condition, the payments scheme, like the "right to pollute," puts threat potential and hence bargaining advantage in the hands of the polluter. (For further discussion of this point, see Kneese and Bower, 1972, especially p. 143.)

Exercises

○ **1** **(a)** In Matrix 6.6 find:
- **(i)** Any dominant strategies.
- **(ii)** Any equilibrium cells.
- **(iii)** Each player's maximin among *pure* strategies.
- **(iv)** The intersection of the strategies in part (iii).

(b) Answer the following:
- **(i)** What can A guarantee himself? How?
- **(ii)** What can B guarantee himself? How?
- **(iii)** What can A keep B down to? How?
- **(iv)** What can B keep A down to? How?

(c) Give the coordinates of:
- **(i)** The Shapley status quo point.
- **(ii)** The Nash status quo point.
- **(iii)** The negotiated outcome, using the Shapley status quo point.
- **(iv)** The negotiated outcome, using the Nash status quo point.

(d) Draw a large joint-payoff graph, labeling the convex hull, the negotiation set, and all the points corresponding to answers in part (c).

○ **2** For the game shown:

0, 0	2, −1
1, 2	3, 1

(a) Draw the payoff polygon in joint-payoff space.

(b) Does A have a "threat strategy" that coincides with his "security strategy"? If so, name it.

(c) Same as part (b), but for player B.

(d) Is it an advantage or a disadvantage to have one's threat strategy and security strategy coincide? Why?

(e) How does this advantage or disadvantage show up in a method of solution for variable-sum games? (Be brief.)

○ **3** In the game for exercise 2:

(a) Draw the payoff polygon, showing that the negotiation set is not inclined at 45° to the axes.

(b) Find the Shapley status quo point.

(c) Use the method of Graph 6.5 to go from the Shapley status quo point to a solution.

(d) Find the equilibrium point.

(e) Is this equilibrium point threat-vulnerable in the sense of Section 4.3? If so, who is the threatener?

(f) What proportion of the total possible capitulation is called for by your answer to part (c)?

○ **4** In the pauper-millionaire inheritance situation at the end of section 6.1.2,

(a) Draw the payoff polygon, with the millionaire's utility on the horizontal axis.

(b) Draw a new graph showing only the payoff points (6, 14) and (7, 12) and the line through them, extended indefinitely in both directions.

(c) Using (0, 0) as the status quo point, show how the solution is found, noting which segment of the negotiation set it lies on.

○ **5** Creative acts such as songwriting and scientific research have positive externalities, just as pollution has negative externalities. And just as an unregulated free-enterprise system would not bring the polluter's ill effects into his production decisions, it would likewise fail to provide much encouragement to scientists and songwriters. In this light, briefly discuss the significance of copyright, patents, photocopying, tape decks, and the laws concerning these things.

○ **6** It has been observed that if someone in the position of Sufferer in Game 6.2 goes to Polluter to try to bargain, then Polluter can make the pollution even worse in order to have a good bargaining position. Or polluter might threaten to do so. Sufferer, foreseeing these possibilities, might therefore choose to "let sleeping dogs lie." Discuss these possibilities as modifications of Graph 6.8 or 6.10.

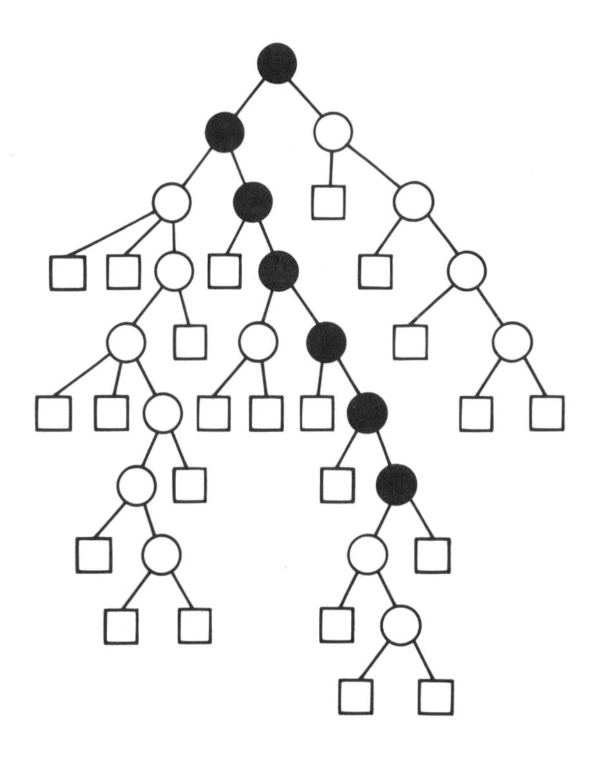

7 Economic and Social Dilemmas

Pollution, city financing, and nuclear proliferation are just a few of a long list of urgent problems that can usefully be analyzed as "dilemma games." The decision-makers in such situations range all the way from individuals to corporations, cities, and nations. As indicated in the first chapter, we shall not leave individuals to the psychologist, corporations to the economist, and cities to the political scientist. Instead, by focusing on interactive-decision structures, we shall find broad similarities among superficially diverse situations. Some differences will also emerge, but these will be based on payoff functions, not on subject matter.

Consider a game with the following properties:

☐ **1** Each player has a dominant strategy.
☐ **2** If each player uses his dominant strategy, the result is worse for each of them than if they all had not.

In short, dominant strategies intersect in a deficient outcome. These two properties were originally introduced as the essence of the two-person "Prisoner's Dilemma." They are, in addition, perfectly meaningful in games with three, four, five, or a million players. The payoff structures of such multi-person "Prisoner's Dilemma" crop up in a wide variety of real situations. Just how wide a variety will become clear in Section 7.2, which contains descriptive analysis of a large number of examples at many levels of social organization.*

To set the stage properly, Section 7.1 presents a formal framework that will allow us to compare these diverse situations. A new kind of payoff graph will provide insight into what motivations may be present for players. Use of the graph imposes two restrictions: the games will be symmetric and each player will have exactly two alternatives. It will not be required that one of those strategies dominate the other; in fact, the various no-dominance configurations are both applicable to real situations and strategically interesting.

The final section of the chapter deals with a variety of techniques for inducing cooperation. These include social, political, economic, and legal maneuvers. They are organized into two broad areas: informational and structural.

7.1 Graphs of Symmetric Dilemma Games

A new kind of graph, introduced in this section, makes it possible to see at a glance the broad strategic properties of a symmetric dilemma game. The first two examples elucidate the approach and allow discussion of the important effects of group size. In keeping with our structural orientation, the subsections are organized by graph shapes—parallel straight lines, curves, and Xs are discussed in turn. The shape of the curves will tell us if it is crucial to get everyone cooperating or whether a few cooperators will suffice. The X-graphs represent situations with different kinds of equilibria and show when a "critical mass" is important.

* For further analysis, see Hamburger (1973b) and Schelling (1973).

7.1.1 Parallel Straight-Line Graphs

● **Game 7.1 Playground**
Three families want to add some equipment to a small playground that adjoins their three lots. The city council refuses to provide funds but allows them to go ahead on their own. The council also points out that it is traditional to have $30 as the standard contribution, for those willing to give. It is assumed to be a general rule of thumb that equipment in a public park gets twice as much use as the same equipment in a private yard.

The families here are players in a three-person contribution game. The standardized contribution and the rule of thumb for toy usage are simplifying assumptions that make it possible to express the payoff structure as a graph. However, once the graph is understood, it will be clear that the general properties of the situation are relatively unaffected by any relaxation of these assumptions. Before going further, the assumptions should be discussed.

To begin with, the three families may have different-size bank accounts and different attitudes that make their utilities different from each other. We ignore such differences and simply express all possible payoffs in the same dollar amounts for each player. This simplification includes our assumption that there is a standard-size contribution, specifically, $30. Thus each family has exactly two alternatives: contribute or not. In this simplified form the situation is, as promised in the introduction to this chapter, a symmetric two-alternative game.

The rule of thumb in the game description reflects the notion that public equipment may expect to be more heavily used than the same equipment in private use, before it rusts out or is outgrown, because it is accessible to more people. Thus, money spent on it achieves more total usefulness or value when spent publicly than it would privately. Specifically, we are supposing that every single dollar spent on park equipment provides two dollars worth of pleasure to the community (of three families) during the useful life of the equipment. We will thus say that the "group-use factor" is 2. The particular choice of 2 is somewhat arbitrary but has the virtue of lying between 1 and 3. The significance of being greater than 1 is that $1 is what a privately spent dollar is worth and the whole point of a community project is that there are joint benefits possible from joint action. On the other hand, a value greater than 3 would suggest that *each* family got more use out of the jointly owned equipment than it would from the same item privately owned. Note that we are basing value on use here, on the assumption that privately owned things are not shared (if they are, then they are "semipublic") and we are also ignoring any value of social interaction (kids and parents meeting informally in the park).

In sum, each family may contribute to this project or not, and the value of the equipment ultimately purchased will depend on total contributions. This total is multiplied by the group-use factor of 2 and then the resulting value is divided equally among the three families. For example, if all three families give $30 each, a sum of $90 will be multiplied by 2 to yield $180 worth of community value. Split three ways this gives $60, so if all families give they each double their pleasure, from $30 to $60. This is a net gain of $30. This result is shown in the top line of Table 7.1. The next line of the table shows

Table 7.1 Summary of the calculation of payoffs for "Playground."

	Number of families	Cost to each	Total gifts	Value to community	Benefit to each	Net gain to each
Giver	3	$30	$90	$180	$60	$30
Giver	2	$30	$60	$120	$40	$10
Non-giver	1	0				$40
Giver	1	$30	$30	$ 60	$20	−$10
Non-giver	2	0				$20
Non-giver	3	0	0	0		0

what happens when only two families give. A total of $60 is then given, having a community value of $120 (2 × $60) for a gain to each of $40 minus whatever each gave. For givers the net gain is $10 ($40 − $30) while for nongivers the $40 is pure gain. This $40 is the highest net gain or payoff possible anywhere in the game, as can be seen by looking at the right column of the table. Thus, the best result occurs for the single family that does not give when the other two do give. Similarly, the worst result, −$10, befalls the "lone giver" (this reckoning ignores any satisfaction that may derive from the charitableness of the action).

It is possible to construct a three-dimensional matrix, cubical in shape, to represent this situation. The interested reader may enjoy contemplating Matrix 7.1, in which C stands for "contribute" and D stands for "don't contribute." In addition to two rows and two columns, the matrix has two levels. Three-dimensional matrices are awkward enough, but they are as nothing compared to the four-, five-, and higher-dimensional matrices that would be needed if there were more players involved. It will therefore be wise for us to turn to

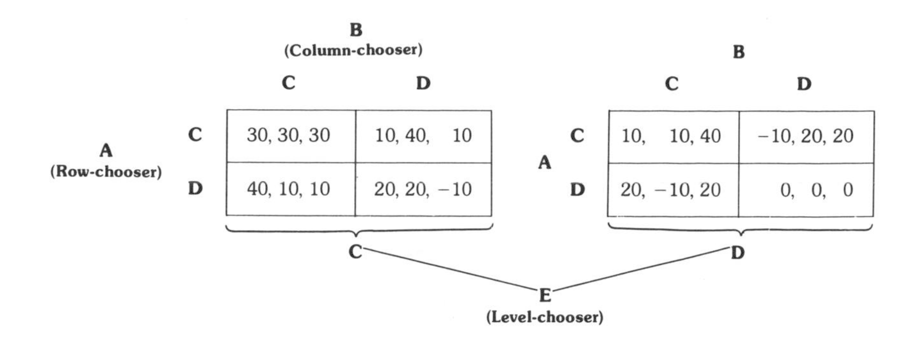

Level C:

A \ B	C	D
C	30, 30, 30	10, 40, 10
D	40, 10, 10	20, 20, −10

Level D:

A \ B	C	D
C	10, 10, 40	−10, 20, 20
D	20, −10, 20	0, 0, 0

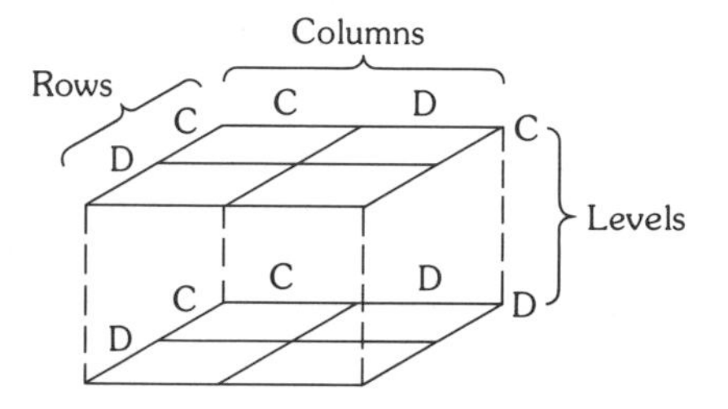

Matrix 7.1 A three-dimensional matrix for "Playground." Such matrices are awkward and will not be used.

a different representation of games of this sort. For symmetrical two-alternative games it turns out that payoff information for large numbers of players can be represented clearly and conveniently in a simple graph.

Graph 7.1a shows what happens if a player picks alternative C. That player receives 30 if the two others both also pick C, only 10 if one of them picks it, and −10 if neither of them does. These num-

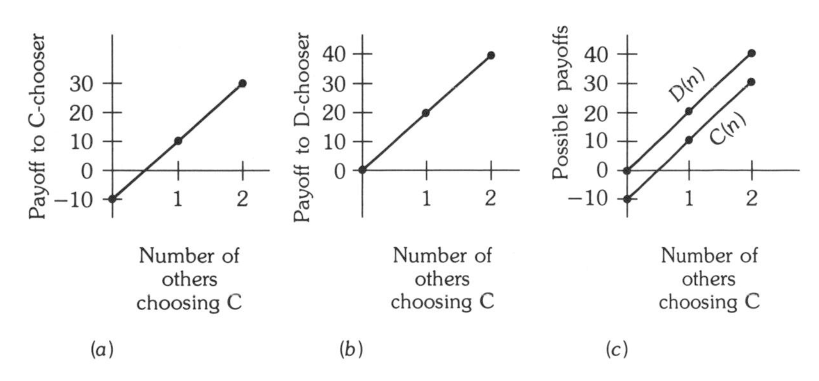

Graph 7.1 "Playground" payoffs for a player who picks C—Graph (*a*)—or D—Graph (*b*). Graphs (*a*) and (*b*) are superimposed in Graph (*c*).

bers can be checked against the last column of Table 7.1. Similarly, Graph 7.1b shows the possible results for a player who picks D, but the horizontal axis is still the number of players picking C (not D). The largest possible payoff, 40, occurs for a player who picks D while both others pick C. This and the other values on the graph are found from the table.

It will be convenient to refer to these graphs using the notation of functions. For example, C(2) stands for the payoff to someone who picks C when two others also pick C. The number in parentheses always refers to the number of others who pick C. Thus, D(2) is the payoff to someone who picks D when two others pick C (not D). Graph 7.1a is the graph of the C function, showing the values of C(0), C(1), and C(2). Similarly, the D function appears in Graph 7.1b.

These two graphs are superimposed for convenience of comparison to form Graph 7.1c. To see that this graph is a "Prisoner's Dilemma" graph, notice two things about it. First, D will be the dominant strategy if you are better off with D no matter what the others do. The graph has this property since $D(n)$ is above $C(n)$ all the way across; more formally, for any n, it is true that $D(n) > C(n)$. Since the game is symmetric and the graph represents the view of any player, D is dominant for everyone.

In addition to dominance the graph also shows the other essential aspect of "Prisoner's Dilemma," namely deficiency of the equilibrium. This is because the right end of the C graph is higher than the left end of the D graph, i.e., $C(2) > D(0)$. Notice that C(2) is the payoff to someone who picks C when two others also pick it, i.e., the payoff when all three cooperate. On the other hand, in D(0) the zero means no others cooperate so D(0) is the payoff when all choose D. In numerical terms, if everyone chooses D then all of them get 0, whereas if all choose C they all get +30. Thus, the all-D outcome, resulting from dominant, "individually rational" choices, is "deficient" in that all players could simultaneously do better by doing something else.

This kind of graphical representation is easily extended to games with any number of players. For example, a contribution game for 10 players with a "group-use factor" of 4 is shown in Graph 7.2. It is plausible that a larger group-use factor could be used for a larger group since there are more potential users of the community property. Nevertheless, the group-use factor chosen here is not as large *proportionately* as that for the three-player game; this reflects the idea that a park cannot be as readily accessible to all members of a larger group.

The extent of cooperation in contribution games such as these depends on several factors, including the number of players and the

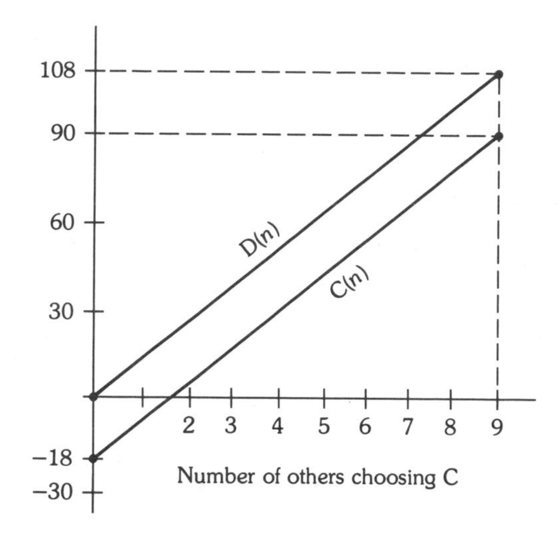

Graph 7.2 A 10-player version of "Playground" (C = contribution; D = no contribution).

payoff structure. A larger group will tend to have relatively greater difficulty in communicating, and in reaching and enforcing agreements. Even in the absence of communication, small groups have the inherent advantage that each player is a significant fraction of each other player's environment. The extreme case is the two-person game where you are the *only* player seen by your counterpart.

Payoffs can make a game easy or hard to achieve cooperation in, whatever the number of players. The effects of payoffs, however, interact in interesting ways with the number of players. To see some of the possibilities, we shall now compare the payoff structures of Graphs 7.1c and 7.2.

In each of the two games described above, there is a fixed sacrifice involved in contributing—$10 in the first game, $18 in the second. Suppose you decide to make that sacrifice, hoping that you will influence some others to join you; we now ask how many others would have to be so influenced in order for you to make a net gain. In the first game your payoff goes up $20 with each decision by another person to contribute. Therefore, you need influence only one other person in order for your move to C to yield a net gain. Thus, C(1) exceeds D(0), and C(2) exceeds D(1). In the second game contributions by others yield you a value of only $12 each, so that two others must be influenced if your move to C is to yield a net gain. Thus, C(2) exceeds D(0), though C(1) does not exceed D(0).

The foregoing considerations might seem to suggest that contribution is more likely to be rewarding in the first game where only one other person need be influenced. However, notice that this means half the available other players, while in the second game a player's contribution only has to influence 2/9 of the others, that is, two out of the nine. Also notice that the potential gain in going from no contributions at all to all players contributing is five times the size of a player's sacrifice ($C(9) - D(0) = 90 = 5 \times 18$), while the corresponding ratio is only three in the first game ($C(2) - D(0) = 30 = 3 \times 10$). Experiments conducted with various numbers of players are reported in Section 9.5. The results indicate that, other things being equal, cooperation is relatively less frequent in large groups. Moreover, it becomes almost impossible to coordinate group-wide cooperation with more than three people and no communication.

7.1.2 Curves, Xs, and Other Graph Shapes

In both of the above contribution games $C(n)$ and $D(n)$ are straight lines and are parallel to each other. This property of the graphs reflects the fact that each of these games is "decomposable" or "separable," in the sense of Section 4.2.2. This separability results from our assumption that *each* dollar yields a resulting value to the group of two dollars worth of usefulness (four dollars worth in the second game). Other assumptions are possible and we shall now look at a couple of possibilities.

Suppose that beyond a certain point additional contributions to the park fund are just icing on the cake, since the key items can be purchased even with a small budget. Graph 7.3 shows that in this

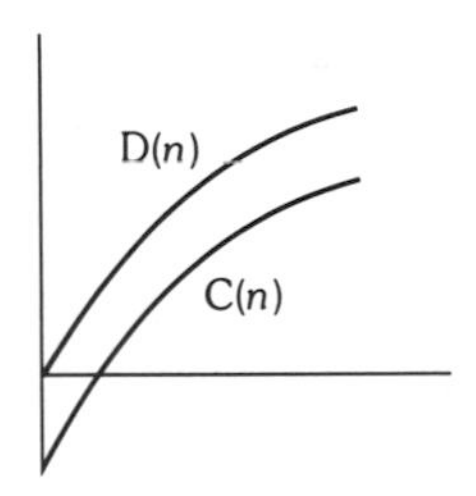

Graph 7.3 A contribution game in which the first few contributions are more important than the last few.

case the benefits rise less steeply as more players use C. Another alternative would have been a threshold model, with the assumption that below a certain number of contributions there would be no hope of making the park enticing enough that people would bother to come. Thus, in Graph 7.4 the payoff functions are flat on the left side, in the region of a below-threshold number of contributors. In both Graphs 7.3 and 7.4 the difference in height between $D(n)$ and

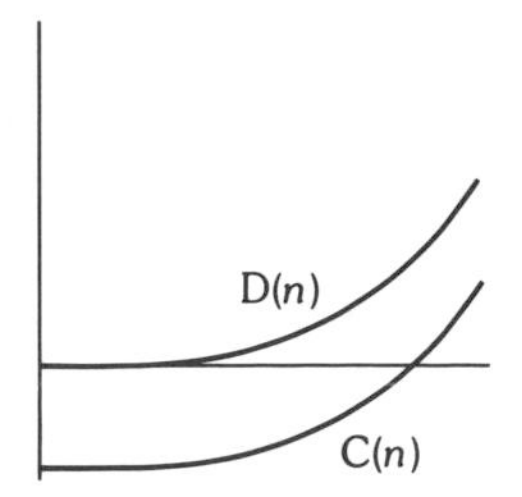

Graph 7.4 A contribution game in which contributions are of little use if there are too few of them.

C(n) is a fixed amount all the way across; the difference equals the amount of the contribution.

"Blackout" is played by the population of New York City on hot summer days when the mighty generators cannot quite manage to satisfy the residents' combined appetite for air conditioning. This game, though superficially very different from financing a children's park, turns out also to be a "Prisoner's Dilemma." Graph 7.5 is proposed as the payoff structure of this game. Let us see why.

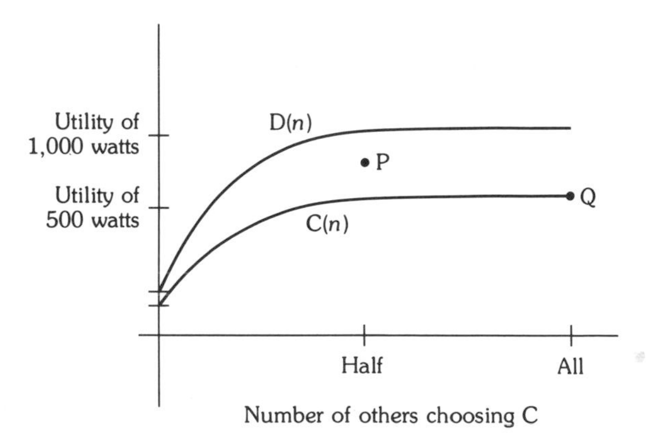

Graph 7.5 "Blackout" (C = not use air conditioning, D = use air conditioning).

Suppose that each household is deciding whether to use 500 or 1,000 watts of electricity. It is assumed that every household normally uses 500, for a refrigerator and a few light bulbs. The second 500 is the discretionary amount. The "C" strategy here is self-abnegation, going along with the mayor's broadcast plea not to use air conditioning. The tricky part of this particular game is that 50 percent (or more) of the people might as well be comfortable since there is really no strain on the power system until 80–90 percent of the people use air conditioning. But which 50 percent? Sweltering in your living room, listening to the mayor's exhortations (on a battery-operated radio), knowing that most of your fellow-residents are not following your example and are not even aware of your sacrifice, suspecting moreover that one teeny-weeny extra 500 watts among the billions cannot possibly be the proverbial straw that breaks the camel's back. ... What do you do? Why of course you turn on your air conditioner—just like everybody else. And what do you know, surprise of surprises, five minutes later there is a blackout.

The right side of Graph 7.5 shows that as long as a substantial number of people refrain from using the extra power there will be virtually no chance of a blackout. In that region D-choosers get their 1,000 watts with certainty and C-choosers get their 500. The first 500 is for essentials, while the additional 500 is only for comfort, so the C graph is more than half as high as the D graph. Moving leftward, as fewer and fewer people refrain, that is, as more and more turn on their air conditioners, the likelihood of a blackout becomes greater. This is reflected in a drop (as we move left on the graph) in *expected* utility. If no one refrains (everyone uses 1,000 watts) then the probability of a blackout is high, say 3/4. If the utility of a D choice was U when there was no chance of a blackout, it now becomes $1/4(U) + 3/4(0) = 1/4U$. A C-chooser's payoff is similarly reduced. Thus, at the left side of the graph (no refrainers) the graphs are 1/4 as high as on the right (all refrainers).

The graph of "Blackout" bears a certain resemblance to the graph of the "icing-on-the-cake" version of the contribution game (Graph 7.3). In each case the best "return on investment" of C choice is obtained when only a small number of players choose C. That is, the curves rise more steeply on the left side than on the right. The effect is more severe in "Blackout," where there is no increase in benefit at all beyond a certain level of cooperation.

To make the point differently, notice the points P and Q in Graph 7.5. The height of Q is the average payoff when all refrain from using air conditioning. In this case the averaging is trivial because everybody gets the same result. P is also an average payoff. When half the people pick C and half D, then the average payoff is midway between the two graphs, as shown by the position of P. Other averaged values could be shown, but these two suffice to make an important point: since P is higher than Q, society is, on average, better off at P than at Q. Half the people can be comfortable at no cost to the others. But which half? Sitting alone in your living room, knowing that it *makes sense* for half (or more) of the air conditioners to be turned on, you decide that yours will be in that half. This kind of reasoning may seem somewhat group-oriented, but the generator does not know how you reached your decision!

If millions of people could communicate effectively, perhaps they could agree to take turns. It may be possible to decide *for* them by imposing "rolling blackouts," deliberately blacking out one region of the city after another. This of course puts the affected area down not just to 500 watts but to 0. This has been done in New York and in 1977 was proposed as a standby plan in the Pacific Northwest and in Northern California, where a long drought threatened hydroelectric sources.

In "Littering" each individual decides whether or not to litter. However, the trashcan slogan "Every litter bit hurts" is not a correct reflection of the payoffs, according to Schelling (1971). For him "it is really the first few litter bits" that do damage.* In other words, payoff drops off steeply as the number of cooperators drops from universal cooperation at the right side of the graph. Moving leftward, i.e., as fewer cooperate, there comes a point after which no more significant damage can be done, since the environment is already spoiled. The left side of the graph is thus flat, and we conclude that "Littering," in Schelling's view, has roughly the same payoff structure as the threshold model of the contribution game in Graph 7.4.

Congestion can occur at parks, beaches, and the roads leading to them, on city streets, in restaurants and restrooms, tennis courts and telephone booths. Like the blackout, congestion results from too many people trying to do the same thing in the same place at the same time. Congestion games are particularly appropriate for the kind of payoff graph used in this chapter because, unlike contribution games, they typically present just two choices—either you go to the beach or you do not (for other either-or situations, see Schelling, 1973). On the other hand, although there is some kind of dilemma involved, these situations do not have the structure that defines "Prisoner's Dilemma" since they do not have a dominant strategy.

Congestion typically occurs only at certain fairly predictable times; beaches and parks, for example, are most heavily used on summer weekends. At other times the congestion game, if it can be said to exist at all, is not of interest, so we restrict attention to known peak-load periods.

Turning to the payoff structure of a specific congestion game, suppose that you are deciding whether to spend the Fourth of July at Yellowstone National Park. If you stay home you may never know (and presumably do not care) whether the park was jammed or not. That is, a nonuser's payoff is unaffected by the value of n, and is therefore horizontal in Graph 7.6. The user, on the other hand, is

---

* In addition to making this whimsical observation, Schelling has done more than any other social scientist to meaningfully apply the conceptual framework of game theory in the analysis of political, social, economic, and international situations.

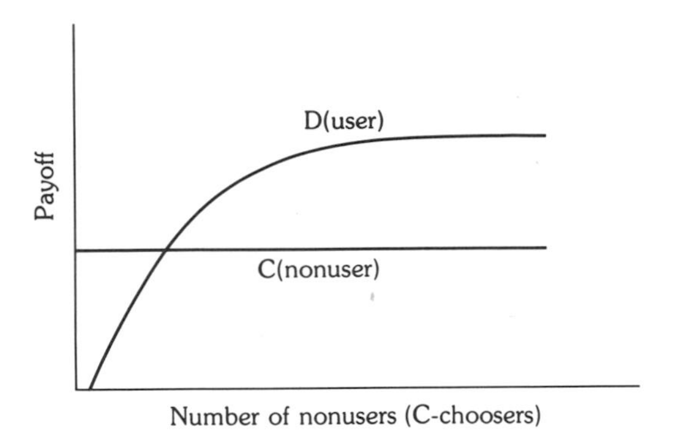

Graph 7.6 A congestion game.

very much affected by how many others join him. If everyone comes he will wish he stayed home, while if the turnout is below some reasonable level he will be glad to have come. Of course, he does not need the whole park to himself, so part of the user's payoff function is flat, too. Since neither option is always the better one, there is no dominant strategy and hence no "Prisoner's Dilemma." However, it does seem appropriate to call this payoff structure a dilemma game of some sort, perhaps "Congester's Dilemma."

In "Tax Flight" large businesses decide whether to move their corporate headquarters away from their present location in the downtown area of a city and relocate where taxes are lower. (There may also be other reasons for moving or staying.) With each business that flees, the tax base is diminished by a certain amount, and those that remain face either higher taxes or reduced services unless some new company comes in to replace the old one. Graph 7.7 portrays a situation in which if all the businesses currently in some city were to remain there, then none would have a reason to regret doing so,

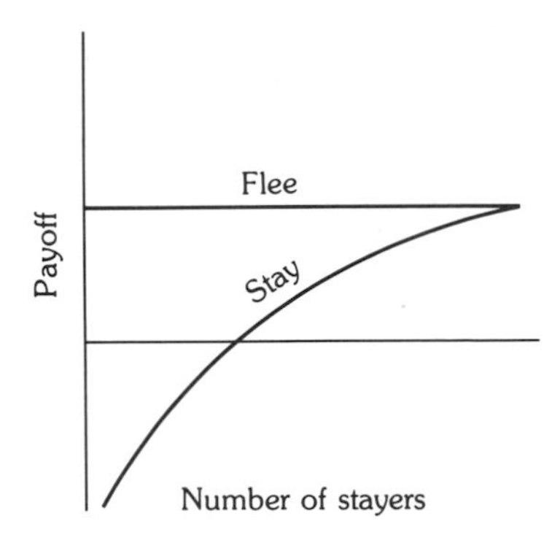

Graph 7.7 "Tax Flight."

since payoff is the same for C and D at the right side of the graph. This is a very shaky equilibrium, however, since if any one of them moves then all the others will want to move, and with each additional move the incentive for additional ones will grow stronger. This graph thus represents a situation with a potential avalanche of decision-change.

This game is not so much a dilemma for those who play it as it is for the city and the individuals who live in it and who may not be able to move at will. Like the hostages in a prison riot, these people have preferences among outcomes but do not have decisions to make that will influence which outcome occurs. For the players themselves the decision to move has a constant payoff; once a business moves from city to suburb it is no longer much affected by how many others also leave the city. Thus, the "uncooperative" equilibrium is not deficient. City administrations on the other hand are indeed affected by the results of these moves and many of them have set up special offices to root out rumors of moves and to entice or cajole companies into staying.

"Backpatting," the final game in this section, may look facetious but it incorporates a very real notion of social solidarity and is intended in all seriousness. It is simply the contribution game with a new wrinkle. "Backpatting," however, has quite different strategic properties. Suppose that each contributor gets a little button to pin on his lapel, saying "I gave." Then whenever two people notice each other wearing these buttons they can pat each other on the back (or exchange self-righteous smirks) and feel good about it. If one assumes that there is no limit to how many pats on the back a person can exchange and still take pleasure from additional ones, then the more contributors there are the better off each will be.

To take a numerical example, suppose that the game of Graph 7.2 is altered by adding a "backpatting factor"; each contributor gets an amount of backpatting pleasure equivalent to $3 for each other individual that also contributes. Thus, if you are the only contributor you will have no one to pat backs with, so C(0) is the same as in the original game. On the other hand, if everyone contributes, everyone will have nine other backs to pat for a total of $27 additional, giving $C(9) = 90 + 27 = 117$. All of the payoffs to D-choosers will be unchanged from Graph 7.2. The resulting payoffs are shown in Graph 7.8.

Notice that the D graph is not everywhere higher than C; D is thus not a dominant strategy and the game is not "Prisoner's Dilemma," though it does have a deficient equilibrium. However, it also has another equilibrium in which everyone gets the best available payoff. These assertions will be supported in the following discussion of equilibria of the last three games.

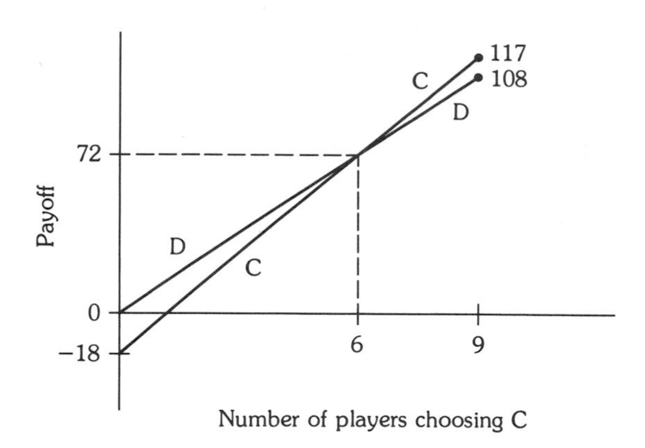

Graph 7.8 "Backpatting."

Congestion games, "Tax Flight," and "Backpatting," as we have presented them, all have intersection points on their graphs, points where the C and D functions take on the same value. Each such point is by definition an equilibrium since no single individual can increase her payoff by unilaterally changing her choice. Yet as noted for "Tax Flight" in Graph 7.7, such an equilibrium may be highly unstable.

Examination of the criss-cross Graphs 7.6 and 7.8 will show that their equilibrium properties are very different from each other. Looking first at the right side of Graph 7.8, notice that when everyone contributes, everyone gets maximum payoff and no one is tempted to switch. This point is not only an equilibrium, it is a stable one. Thus, suppose there were a small number of noncontributors, sufficiently few so that we are still to the right of the crossing-point of the C and D graphs. In this region C gives the higher payoff, so the tendency is to stay with C or switch to it. Since the horizontal axis measures the number of C-choosers, the incentive to pick C in this region tends to move the outcome rightward on the graph, back toward universal contribution. On the other hand, if there are so few contributors that the outcome is to the left of the intersection of the graphs, then D is above C and there will be a tendency to pick D, moving the outcome even further left. In summary, the outcomes will tend to move away from the intersection point in Graph 7.8 toward one or the other of the two outcomes where everyone does the same thing.

The congestion game of Graph 7.6 also has a crossover point, but one with entirely different properties. In this case it is the D function that rises more steeply from left to right. Therefore, in the region to the right of the intersection point, D is attractive to players because it has the higher payoff, so there will be a tendency to lose cooperation there, moving the outcome leftward toward the intersection point. On the left side C has the higher payoff, favoring more C choice, which will move the outcome rightward toward the intersection point. This point is thus an equilibrium point. Wherever the outcome is located, there are payoff incentives that induce movement toward that equilibrium.

Returning briefly to "Tax Flight," we may speculate that although the current situation is as represented in Graph 7.7, there was a time when the game looked more like Graph 7.9, in which the

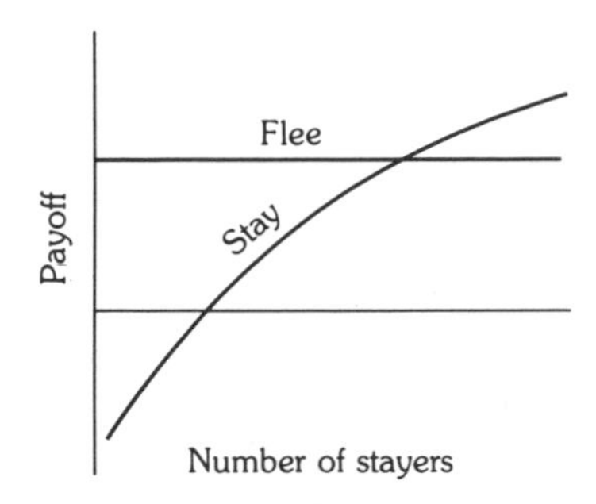

Graph 7.9 "Tax Flight" at a point in time before the situation represented in Graph 7.7.

outcome of everyone staying is an equilibrium. Presumably, events external to the game had the effect of lowering the desirability of staying in the city, shifting the payoff function for those who chose to stay and creating conditions for flight.

7.2 Real-World Dilemmas

Who among us has not indulged himself in a little self-righteous complaining about the sorry state of the world or some part of it? Particularly frustrating, it seems, are situations in which coordination and cooperation are all that are needed to overcome a problem. In this section we shall analyze such problems in terms of the conceptual framework of Section 7.1.

Public indignation has rarely been higher than it was in 1976 when it was revealed that corporations had been paying off foreign governments to obtain contracts or for assistance in violating local business regulations. From princes and prime ministers on down the line, officials were found with their fingers in various cookie jars, proffered by corporation representatives. The most popular excuse offered by corporate officials seemed to be "everybody else is doing it." To anyone who has been caught cheating in school, this excuse will probably have a familiar ring. But even when no one was doing it the temptation was present; after all, the practice had to get started at some time. We do not have to suppose that all morality is lost, only that it is less compelling than this prospect of gaining competitive advantage. With this assumption the situation becomes "Prisoner's Dilemma," since if the corporate representatives all used the dominated strategy of refraining from bribery, none of them would be at a competitive disadvantage and they would all be spared guilty feelings and fear of being caught.

A surprisingly large number of the more maddening social, national, and world problems turn out, on inspection, to be some kind of dilemma game, at least in part. This assertion will be supported by a series of examples in this section. The next section discusses practical methods of resolving social dilemmas and also introduces several more examples.

7.2.1 Two-Tiered Dilemmas

A remarkable thing about pollution as a dilemma game is that it occurs at so many levels of social organization. Individuals, businesses, cities, states, and even nations face a variety of decisions in which it is cheaper and simpler for them individually to pollute more. Yet the sum total of these separate little decisions to pollute is often an end product that all parties would find worthwhile to avoid if only they could reach an agreement, and keep or enforce it.

Take, for example, smog devices for cars. When smog devices were optional very few were sold, even in the face of daily eye-irritation reports and warnings not to let your kids run because deep breathing might be hazardous to their health. An unwillingness to buy the device is individually rational in the sense that one smog device cannot solve the problem. A single device might lower the concentrations of noxious chemicals by perhaps a thousandth of a percent, which in turn might slightly lower eye irritation and the likelihood of lung ailments for each person. But these minuscule changes, viewed as benefits to the single individual deciding whether to use a smog device, may not seem worthwhile when compared to the possible $5 or $10 per month in extra gasoline due to lower performance by the car.

If, on the other hand, the government requires every car owner to have a smog device, then the result is indeed noticeable. It can be detected not only by hospital statistics (fewer emphysema patients admitted) but by the naked eye and nose. Cleaner air is very popular but was not achievable by individual voluntary decision-making. Why? The dilemma arises because so much of the benefit from cleaning up your car's exhaust is, in the terminology of Section 6.2, an external benefit, external to you, the decision-maker. Cleaning up

1/100,000 of the smog may not be much, but that improvement is available not only to you but also to the other 99,999 people in the community. The internal benefit—very slightly better air for one person—affects your decision directly. The much larger external benefit—very slightly better air for 99,999 people—affects your decision only insofar as you may be altruistic (or religiously, culturally, or philosophically motivated, etc.).

This dilemma for the driver creates a complementary dilemma at the level of the car manufacturer. If nobody wants to buy one, then the individual manufacturer who makes or builds in a smog device will be at a disadvantage. Indeed, before standards were legislated only a handful of cars were available with a built-in smog device. On the other hand, if all manufacturers install the device and if prices are correspondingly higher, the result may be higher profits all around (except insofar as higher price may result in slightly lower sales).

There is another pollution dilemma that works on two levels but which is more difficult to solve because the higher-level players are nations. The lower-level players are manufacturers of all sorts, not just car makers. Each steel corporation, for example, faces a variety of decisions concerning the pollution level of its factories. If there are no regulations or taxes on pollution, it will typically be cheaper to pollute more. Again, this dilemma can be solved at the national level, but now a new problem arises. Suppose that a nation does regulate pollution by its steelmakers. Then, when these companies attempt to sell their steel on the international market, they will have to raise prices, cut profits, or both. Thus, a nation with tight pollution laws faces a competitive disadvantage for its products and a resulting balance-of-payments problem. Again, the dilemma is potentially solvable at a higher level. However, when nations are players, the only higher level is the international one, where no government exists. Therefore, it is only through negotiation that international agreements concerning pollution in manufacturing can ultimately be reached.

Two interacting levels of dilemma can also arise when individuals band together in their group interest. Imagine that the group is a national organization of small businessmen lobbying for subsidized loans, or a professional group, say dentists, lobbying to maintain tax deductions for attending conferences. Each individual businessman or dentist must decide whether or not to give his time or money to the effort. No single contributor will significantly affect the degree of success of the lobby, and any legislation that is obtained will benefit noncontributors as well as contributors.

There is another, higher-level, dilemma in which the special-interest group itself is the player. To see the nature of this higher level more clearly it will be helpful to envision an idealized situation in which (1) everybody belongs to exactly one special-interest group; (2) each group is the same size; (3) each group is lobbying for a monetary advantage of the same amount; (4) each lobby will get what it asks for if it works hard; and (5) the government offsets what it gives to special interests by increasing general taxes on everyone. In such a situation, if every interest group chooses to lobby, then everyone will get special benefits and lose an equal amount in increased general taxes. Thus, each group will achieve precisely nothing for its members, at considerable organizational expense in time and money. (One benefit may be less unemployment among lobbyists, but let us suppose that the lobbying is done by members of a group on a shared basis.)

At this higher level the cooperative strategy for each special-interest group is to refrain from lobbying; we have just seen that if every group does this, all are better off. On the other hand, if each group considers only its own benefit, each will do better to lobby, regardless of how many others do so. This is because even though by getting its own special benefit the group causes general taxes to increase for everyone, its share of those taxes is only a fraction of its benefit. The ironic aspect of the two-level dilemma here is that it seems to be self-contradictory. For example, an individual who wanted to be "cooperative" at both levels would join a group that does nothing!

Legislative log-rolling is a two-level "Prisoner's Dilemma" closely resembling competition between special-interest groups. On the lower level, two legislators who agree to vote for each other's pet projects are engaged in a pact that affords joint benefits that otherwise might not be possible. This is a dilemma situation because it is also true that each would prefer not to have to live up to his half of the bargain. At the higher level, if such pacts are common throughout the legislature, the overall result may be a series of narrow-minded projects that, taken together, are not efficient for the state as a whole (for a technical treatment, see Brams, 1975, Chapter 4).

7.2.2 Women, Men, and Couples

When people demand to be treated equally it is usually because they are treated as less than equal—in rights, opportunity, or status. The privileged do not demand equal treatment. If life were a zero-sum game, the gaining of rights by some people would necessitate corresponding losses by others. Even though life in general and struggles for equal rights in particular need not be zero-sum, many people often carelessly perceive such struggles as if they were. Because of this (but also for other reasons) leaders of organizations seeking equality of rights are often labeled "pushy," "strident," and "demanding." Leaders in the feminist or "women's liberation" movement of the 1970s are commonly labeled "unfeminine," presumably reflecting the belief that it is a man's prerogative to act assertively, be competent, and wear pants.

The players of "Women's Lib" are women, each of whom must decide whether or not to abandon the stereotype, that is, to stop acting submissive and incompetent. Men will not be regarded as players, although their behavior is indirectly taken into account insofar as it affects payoffs to women. The dilemma is as follows. If all women were to simultaneously change behavior, a man seeking the comfort of a traditional "feminine" woman would have nowhere to turn, and no individual woman would have to contend with the traditional prejudices of men, either professionally or personally. Nevertheless, in reality, in the absence of such universal change, many women who did change their individual behavior found the response unpleasant.

From the viewpoint of the dilemma game, two other phenomena become more understandable. First, the "support group," in which women come together to discuss problems associated with sexual stereotypes, provides the chance to communicate about the dilemma (among other things). Experiments have shown that small groups tend to cooperate in dilemma games better than large ones and that communication tends to induce cooperation. The second observation that the dilemma helps explain is that leaders of women's groups have in many cases created a climate of opinion in which the woman who is "proud to be a housewife" feels defensive.

> It is a shame that the liberation movement has made remarks belittling motherhood. They are practicing the very discrimination that they are so vehemently dedicated to destroying. After all, they are dedicated to the ideals of freedom of choice and equal opportunity for women.
>
> [Letter to the *Los Angeles Times,* Sept., 1976]

This letter makes its own point but misses the point of the dilemma game. Making a woman feel defensive may seem divisive to the movement, and suggesting what she do with her life may seem meddlesome, but any dilemma game requires joint action if the deficient equilibrium outcome is to be avoided. According to this analysis, anyone who is not part of the solution is part of the problem, i.e., is a noncooperator. It is therefore not surprising to find a lot of pressure being exerted on behalf of the dominated strategy (cooperation). Pressure on potential noncooperators is institutionalized in the case of the "union shop," where by contract with the employer only those who join the union are allowed to work (see Section 7.3.2).

As women become liberated from the home, men may become liberated *for* it. This is no contradiction. For men as well as for women there are stereotypes, and there are social penalties for violating either kind—"effeminate" is no more complimentary than "unfeminine." Liberating men for the home means not stigmatizing them for taking care of young children and participating in what have been known as the "joys of motherhood": washing diapers, cooking, sewing, etc.

The dilemma in the game of "Men's Lib" is suggested by a remark I overheard in the university cafeteria: "But a man can't compete in the world if he spends so much time at home." It is true, of course, that the young male executive, lawyer, or professor who does housework while his wife works at *her* career may find that advancement goes to others who spend more time at the office. But this statement is equally true for the young *female* executive, lawyer, or professor who does housework while her husband pursues *his* career.

To formulate a model of "Men's Lib," imagine a society in which all men are fathers and each one must choose to spend either a lot of time with his family and the minimum time required for his job or else little time with his family and extra time at work. Suppose we observe that every man takes the second option. There are at least three possible explanations for this behavior. The first and simplest is that this is what each prefers. Second, one might imagine that each would prefer to stay home more but that there is direct compensation (overtime pay) that more than offsets the loss of time at home. A third explanation, one that involves a "Prisoner's Dilemma" structure, could be valid even if each man wanted to stay home more and there were no overtime pay. Suppose that promotion is based partly on competence and partly on willingness to work overtime. This incentive may induce everyone to do extra work so that promotions must be based entirely on competence. Ironically, if no one were to work extra, the decision would also be based entirely on competence! Thus, if everyone works extra the result is a deficient equilibrium, since all could do better by simultaneously not working overtime, but no single individual will unilaterally drop out of the competition.

Concern for population growth has a long history, going back at least to Malthus in 1798. The gut issue is whether food supplies can match food requirements, but side issues abound. Even if sufficient food were produced, it might not reach those who need it; this is a question of politics and economics. Food may be produced by techniques that have deleterious side effects, e.g., the use of DDT; this is a question of biochemistry and ecology.

The "green revolution" has multiplied food supplies but population continues to multiply even more. The world's population has a doubling time of under 50 years. If the average couple has three children, population doubles in less than two generations. A population that is doubled five times becomes 32 times as large. Can there be a series of green revolutions to keep up with these increases? The question of food aside, more people ultimately mean less energy and space per person. A doubling rate of 50 years continued for 1,400 years (not that long a time in our cultural history) would result in standing room only, wall-to-wall people filling 1,000-story buildings on every square inch of the world's land mass. Of course this will never happen; war, famine, pestilence, or a lower birthrate will prevent it.

Let us assume that population is indeed a problem, and that in the long run everyone will be far better off if the average birthrate is two children per couple rather than three. We also assume that at least some couples have children by choice, not by chance. Consider a couple deciding whether or not to have an additional child. That single child's effect on the world food supply is insignificant and is not a factor in a family's decision to have a child. Of course, the family's *own* food supply, if for example they are subsistence farmers, may be a factor in their decision, but this factor will be balanced, along with other inconveniences, against the joys of parenthood and the prospects of support in old age.

If, on balance, the couple wants another child, then they might as well have one, in the sense that if the other billion couples in the world are (on average) having lots of kids, then there will be a problem anyway; and if they are not, then there is nothing to worry about. This view, apparently rational for the individual, can lead to the unfavorable group result of unlimited population growth—or population growth ultimately limited by dire events.

> A common misconception is that a species acts in a group-rational way to preserve itself; in fact, each individual organism of the species acts in an individually rational way to maximize the number of offspring it can produce. A fox acts to maximize the number of foxes it can bring forth, and not to maximize the total number of foxes on the planet. In so doing, the fox considers neither the effects of the resultant fox explosion on the other foxes nor on the population of rabbits. Similarly, the rabbit ignores the problem of its own reproduction for the willow population. If the fox is too good a survivor and its offspring are all expert hunters of rabbits, there would shortly

> be many members of the fox family, and shortly thereafter no more rabbits, which in turn would mean the end of the family line of foxes. [Kahan, 1974]

The fox survives as a *species* only because it is an imperfect hunter, and so *individual* foxes die of starvation. The fox has neither the technology of birth control nor the socio-political sophistication to consider a group-rational outcome.

7.2.3 Subgroup Dilemmas

To pursue a game analysis involves, as noted in the first chapter, stating who the players are and what constitutes their payoffs. In dilemma games we wish to show in particular that the payoff structure is such that there is a deficient equilibrium, possibly resulting from dominant strategies. Our choice of players may determine whether we find a dilemma game; that is, by excluding some players from consideration, we may be able to show that a dilemma exists for the remaining subgroup.

The game of "Waiting" requires no description for anyone who has gone to an Academy Award-winning movie the night after the award. More than a theatreful of enthusiasts will line up outside the theatre, long before the film is to start. If each person in line had delayed his arrival by an equal amount, the same people would see the movie and everyone would save time. One might think that the time penalties in this kind of dilemma game could be reduced by letting people buy tickets and then go away, but then the problem reemerges with ticket lines.* In fact, "serving time" on the line is a kind of price that normally insures that only those who care most will see an event (of course, a rich person can secure the services of a stand-in or buy from scalpers). High (monetary) prices could, alternatively, be used to select the audience, but this method would select out the rich, not the dedicated.

Not everyone minds waiting. Those who do mind it but come anyway are engaged in a dilemma game. On the other hand, there is no dilemma at all for those who enjoy the line.

> Noticing—and getting noticed—is what makes waiting to see a popular "in" movie not just another deadening urban plague to be endured but instead a tolerable and, for many, even pleasurable pastime in itself. The longer the line, the younger, more modish it is likely to be, and the more bemused—and

---

* This observation and others that follow (including the quote from the *New York Times*) are taken from a fascinating article by Mann (1973).

> thus not bored—by itself. Almost invariably the predominant conversational gambit has to do with similar evenings at the movies: not remembered great films, but remembered great lines. [*New York Times,* April 25, 1970]

Some people go after a college degree or a trade-school diploma because they like the process of learning or because they will learn something that will be useful in a future job. These people are not part of the dilemma game of "Credentials." However, for some jobs the degree or other credential is in reality a "union-card," an arbitrary job-unrelated criterion used by employers simply to sort out those applicants who cared enough. People engaged in obtaining credentials for no purpose other than to qualify for such jobs are engaged in a dilemma, since if they all refused to undergo the credential process they would all be on a comparable footing in applying for those jobs.

In "Hometown Blackout" local television coverage of a sporting event is prohibited unless the game is sold out. A fan has no difficulty in making a decision if he cares enough about attending the real event to pay the price of a ticket whether the game is televised or not. At the other end of the scale is the TV fan who would not pay to go even if a blackout were certain; this person also has an easy decision. The people of interest to us are those who prefer to stay home provided the game is televised, but would pay to go in case of a blackout. If these people have no information on how many will attend, then when the day of the game arrives they will face a situation in which if they all stay home, each would wish he went. This problem is at least partly overcome by making the final decision on TV coverage a few days in advance of the game.

Neighborhoods can be involved in two kinds of prisoners' dilemma. In one, the dominant strategies lead to unnecessary segregation (Smolensky et al., 1968). In the other, which is relevant to this section on subgroups, the dominant strategy is for an absentee landlord to let his house deteriorate since property value is partly determined by the other houses (Davis and Whinston, 1961). A resident-owner who knows his neighbors has other payoff considerations, so that for him there is no dilemma. (For an introduction to urban economics, see Richardson, 1971).

7.3 Promoting Cooperation

"Cooperation" is a word that carries a good deal of emotional and moral freight. Although we shall continue to use it, it is important to remember that "cooperation" is simply our shorthand way of referring to "the dominated strategy in a dilemma game or real-life situation modeled by one." The following paragraphs provide perspective on our use of this word.

Recall that a model excludes certain individuals in the interest of simplicity or because they have no significant choices to make. These people may be harmed if our players find a cooperative strategy, and so they may not think very highly of such cooperation. A clear example of this is an agreement (possibly tacit) by business competitors to keep prices high. This result is very nice for the businesses, but if we expand the model to include the customers it no longer seems so cooperative. As another example, if all of a candidate's supporters contribute to the campaign, that is a desirable outcome for the group but it is not so desirable from the opposition's viewpoint, and it may not benefit society as a whole. Finally, recall that in the game of "Tax Flight" universal use of the dominant strategy did not harm any of the businesses but was damaging to the city.

Cooperation is often a matter of degree. If it is cooperative to conserve energy, how many lights can I leave on and still be a cooperator? Is it cooperative to have two children? To cut back 5 percent on my driving? To give a pint of blood every seven years? In the graphs of Section 7.1 there were only two possible choices. This restriction was simply a convenience to enable us to make a certain kind of graph, and should not obscure the fact that even if we let players contribute any amount they choose, the dilemma remains. A situation can have the properties of a dilemma game even though each player has more than two options.

Cooperation is different things to different people, and is easier for some than others. The wealthy can afford to contribute more; the healthy can afford to lower their thermostats more; the bus is easier to take if you live near a bus stop; and some people like driving less than others. There may even be people who so enjoy the camaraderie of the bus that the fuel conservation game presents no dilemma to them. However, a dilemma game would still continue among the rest of the population, the subgroup for whom individual and group rationality diverge. Such was the case in the examples of Section 7.2.3.

The ideas in the preceding three paragraphs provide some perspective on what the notion of cooperation may mean. We turn now to the major theme of this section, the question of what promotes or inhibits cooperation, i.e., the "resolution" of dilemma situations. To sort things out a bit, we can divide the factors influencing cooperation into four broad categories, which may be loosely labeled as informational, structural, social, and political-economic. I shall first describe the categories and then discuss each in detail separately. However, social factors will be reformulated, as far as possible, in terms of informational and structural factors. Also, political-economic factors typically work by changing payoffs and allowed choices, both of which come under structure. Thus, we will incorporate everything into the two categories of information and structure.

Under informational factors we include communication, which may be regarded as the passage of information from one party to another. An important distinction can be made between the "broadcast" process, in which information, opinion, or explanation emanates from one source (e.g., a government) to many receivers, and the "diffusion" process, in which information, opinion, or explanation can be readily exchanged among all players. There are still other patterns of communication channels, for example, the hierarchical, in which you can talk to your boss but not to your boss's boss.

Information may be about the structure of the game—payoffs, number of players, etc.—or it may be information as to what the other players are choosing to do within that structure. With respect to payoff structure, a person may know or not know how much it costs to burn a 60-watt bulb four hours a day for a month; may know or not know when fusion will be harnessed, or what fusion is; may know or not know how many spray cans will destroy how much ozone and cause how many cases of skin cancer a year. On the other hand, even if you know the payoffs, you may not know what the other players are doing: how much electricity, how many spray cans, or how many boxes of phosphate detergents an average person uses, or if you are a nation, how many bombers your neighbor is building, buying, or selling.

"Structure" here will refer to payoffs, number of players, and rules of the game. In our discussion of payoffs we will consider the question of symmetry. Because graphical presentation (Section 7.1) is an aid to thinking, it is often useful to treat situations as if they were symmetrical even when they are not. However, the asymmetries themselves directly affect the possibilities for resolution, and so merit separate study. The number of players is an important variable since cooperation is more readily achieved in small groups than large. Thus, techniques for resolution of dilemmas might be based on keeping down the number of players, say by creating barriers to entry into the game. An alternative would be to try to restructure a large group into several smaller ones, or to decentralize decision-making. Rules of the game are also an aspect of structure. Examples include how many times a game is repeated, how it is terminated, whether a decision to cooperate can be made and then withdrawn, and whether a decision can be made contingent on the choices of others ("I'll do it if at least half the others do").

The notion of structure just outlined involves concepts of game theory: payoffs, players, rules. Information, the other broad area so far sketched in this introduction to dilemma-resolution, is also directly related to the game framework. For example, structure may allow a player certain moves, while information may reveal to other players whether those options are being taken. In contrast, our third cluster of factors in cooperation, social forces, appears to have no obvious relation to game theory. Familiarity, love, group loyalty, patriotism, tradition, altruism, principle, and conformity are all social forces that can promote various forms of cooperation. None of them appears to have anything to do with strategic thinking, and in fact one could argue that their existence and potency undermine our analysis of dilemma situations. The argument might run something like this. According to the individually rational approach, no one would ever cooperate in multi-person "Prisoner's Dilemma" situations; yet we observe people giving their time, blood, and used newspapers to Worthy Causes; they do this because their decision processes are distracted by interpersonal irrationalities like love and loyalty, or misdirected by "false" decision principles like altruism and tradition; since game analysis is often wrong about the way people make decisions and often wrong about the ultimate choices they make, it is not a useful methodology.

How can a game theorist counter such an argument? The most straightforward response is to try to incorporate anything that affects choice into utility. Thus, in considering my income (M) and yours (Y), I can say, for example, that my utility is $3/4M + 1/4Y$. Such a utility function would mean, for example, that I prefer a situation in which we both make $20,000 to one in which you make nothing while I get $25,000, which in turn I would prefer to having us both make $15,000. This utility function would be somewhat, but not absolutely, altruistic. The other social forces mentioned above might also be similarly incorporated into individual utility functions. We have already seen, in the game of "Backpatting" in Section 7.1.2, how conformity can alter the payoff graph in a particularly interesting way, creating a second equilibrium. Thus, the social factors conformity and altruism are at least partly convertible into utility. Two other social factors, familiarity and group loyalty, play a role in the fund-raising game of Section 7.3.1, below.

Political-economic factors, the fourth and final category, include governmental bans, regulations, taxes, incentives, and direct actions. These may be designed to promote conservation, livable cities, and community health or to inhibit pollution, congestion, and TV violence. Many aspects of various governmental actions have already been discussed in Section 6.2. These will be reformulated for dilemma games in this section. At the international level there can be agreements on trade, pollution, maritime rights, and the testing, construction, and sale of armaments. Enforceability, which we examined in Section 4.1, is a crucial aspect of both domestic legislation and international agreements.

Political-economic factors, unlike the other three types, are not intrinsic to situations but are, so to speak, manipulations of situations "from the outside" by formally constituted institutions. These "outside" formal institutions are nothing but governments, consisting of legislators and chief executives. These decision-makers in turn may be thought of as players in a higher-level game. But this new level must wait for the next chapter.

7.3.1 Information

Let us now watch someone in action, getting others to cooperate. Tom Wolfe (1970) describes a man named Quat getting contributions for the Black Panthers at the New York apartment of conductor Leonard Bernstein:

> ... Now let's start this off with the gifts in four figures. Who is ready to make a contribution of a thousand dollars or more?
>
> All at once—nothing. But the little gray man sitting next to Felicia, the gray man with the sideburns, pops up and hands a piece of paper to Quat and says: "Mr. Clarence Jones asked me to say—he couldn't be here, but he's contributing $7,500...
>
> Then the voice of Lenny from the back of the room: "... I'll give my fee for the next performance... I *hope* that will be four figures!"
>
> Things are moving again ...
>
> [another contribution]
>
> Quat says: "I can't assure you that it's tax deductible." ... These words are magic in the age of Radical Chic: it's *not* tax deductible.
>
> The contributions start coming faster, only $250 or $300 at a clip, but faster ...
>
> [a few witty gibes, and then]
>
> "I know you want to get to the question period, but I know there's more gold in this mine. I think we've reached the point where we can pass the blank checks."

Three things are worth noticing about this and other fund-raising events: (1) everybody can *see* everybody else; (2) everybody can *be seen* by everybody else; and (3) everybody knows why they are all there. There is a general expectation as to what the outcome is to be, and everybody has information about how well things are moving toward it.

A person who sees others giving may be moved to follow suit for several reasons, perhaps a sense of community, a desire to be part of something successful, or a desire to feel at least as helpful as some other person who just gave. If a person is affected in any of the ways just mentioned, then she can imagine that others may be similarly affected by her action of contributing. Thus, she may see herself as continuing a bandwagon effect in which her contribution will enrich the cause indirectly as well as directly. Leverage of this sort reduces the severity of the dilemma. In addition to this imagined leverage that may result from being seen, a person who is seen may feel embarrassed not to give, especially after many others have done so, so that giving has been established as a group norm.

The episode in the Bernstein living room as recounted by Wolfe is a masterful job of insuring that considerations like these have their opportunity to come into play. Clearly, the first two contributions have been planted, to get things started. No one wants to go first and, thank heavens, nobody has to. The wealthy have their chance for public recognition, a partial compensation for the great sacrifice. If you are going to give a large sum, you may as well get public credit for it. By now the norm of giving is clearly established, but the smaller contributors are out of their depth: who can proudly announce a $50 contribution after the $200 and $300 offers have been flowing in? So a new peer group is created—those for whom the blank checks are appropriate. Each person has the semiprivacy of the personal check, where only the person standing next to him may possibly get a glimpse of what he is giving. Perfect! Because the person standing next to you is probably someone you know, someone with comparable income, the one person in the room you do not want to look cheap in front of, and who you may hope to influence by your generosity.

To see and to be seen—that is what one achieves at a fund-raiser, be it with the social luminaries described by Wolfe or at a humble local political gathering. Those who come both see and are seen; those who do not come do not see and are not seen. There are dilemma games where everyone sees equally, others where moves are made in the dark, and still others where the overall level of cooperation is public knowledge but nobody knows who did how much. Each real situation has its own characteristic information conditions and these can affect the resulting levels of cooperation.

The term "behavior information" will be used to refer to knowledge of what other players are doing. Such knowledge is quite different from structural information, such as how much sewage a lake can dissipate by natural processes or what the community chest does with your money, or operating information like how to find riders for a carpool. The following examples show the degrees of behavior information.

The State of California offers personalized license plates to car owners in the state as an incentive to contribute $25 to the state's pollution cleanup efforts. Leaving aside the question of whether the revenues involved are even remotely commensurate to the stated objective, one may note that the game has an advantageous behavior information structure. Specifically, everyone's decision is on display for friends, neighbors, co-workers, and all the world to see for 365 days.

At the opposite end of the behavior information spectrum is the situation created by the Phantom Snow-Sweeper of Ann Arbor, Michigan. In the late fall in the late 1960s the Phantom would send cards asking for payment in advance for one winter's worth of snow-free sidewalks. Those who responded found that on every snowy morning the Phantom had faithfully fulfilled his bargain. But on those same streets the nonrespondents were favored by a free sweep (the Phantom had a motorized snow-sweeper and it was easier to do a whole street than to stop and start). No one ever knew which of his neighbors did what—except, of course, for the Phantom.

An intermediate case of behavior information occurs on the bus. Riders (cooperators) know who is riding and who is not, but nonriders know nothing of such things. The same comment applies to those who do (do not) attend a fund-raiser.*

In some situations information is naturally, unavoidably available to all. In the population game, where the alternatives are producing more or fewer children, there is no way, short of closeting the children, to keep one's decision secret. In other situations, although information is not intrinsic, publication is readily accomplished. The Phantom falls into this category, and indeed he might have been well advised to send a follow-up note with information about who paid

* Insofar as the bus represents a net sacrifice by its riders in the public interest of energy conservation, it may be regarded as a multi-person "Prisoner's Dilemma," with riding as the cooperative alternative. Many caveats are needed here. Some ride the bus out of necessity or because they prefer it; costs and benefits are many-faceted, including money, time, stress, etc.; and more fuel-efficient cars may undermine any energy savings of the bus.

the bill. Some churches publish detailed contribution lists, presumably because they find that it increases contributions. In still other situations publication of individual choices may be impractical. The telephone company can hardly contemplate publishing a list of all those who have cooperatively refrained from unnecessarily using the information operator, but at one point it went on television to make encouraging remarks like "You've made a start..." toward refraining more often.

Wherever the intermediate case applies, where information dissemination is possible but not inherent in the situation, it is worth considering what the consequences of dissemination would be. In some cases it may mean an invasion of privacy or unacceptable pressure toward conformity. Concerns such as these have been voiced by professors with respect to grade inflation. That is, one might suppose that professors would be less likely to give grades far out of line with campus policy or current practice if every professor's grading average were published. The objection that has been raised is that pressure to conform in grading is counter to the aims of the university.

Dissemination may be aimed at potential cooperators, as when a new and growing union publicly lists its membership for all to see. Meetings bring cooperators face to face, but those who do not attend receive no information, so that other means are needed if they too are to be informed.

The pre-canvass and post-canvass meetings of some political campaigns have served an important and often unappreciated function along these lines. George McGovern's 1972 primary campaign used door-to-door canvassing to an unprecedented level. The ostensible purpose of the pre-canvass meeting was to distribute materials and instructions, that of the post-canvass meeting to return materials and report results. But none of these functions required 10 or 50 people to meet together at a prearranged hour. I believe that an unspoken, perhaps not understood, purpose of these meetings was to confirm at the outset the existence of a large cooperative group, to make an implicit covenant, and to return later to confirm that the cooperative strategy had indeed been carried out.

So far the discussion and examples have stressed behavior information rather than payoff information, and the diffusion of that information among all or many players rather than broadcast from one source. Broadcasting is perhaps most effective when done by a president. Roosevelt asked us to put our money back in the bank and Carter wants us to drive less. Roosevelt's appeal had the advantage that the run-on-the-bank game is not really a "Prisoner's Dilemma," since once a certain number of people entrust their money it becomes advantageous to others to join in. Thus, if each and every person believes he will be better off with his money in the bank, then everyone will be right. The people will have jointly made a self-fulfilling prophecy. The payoff graph of the bank-deposit game, like Graph 7.8, has two equilibrium points, and Roosevelt sought to move the outcome from one to the other.

Carter, in contrast, is trying to cajole us into a truly dominated choice. If everyone else curtails his driving, there is no reason why I should, since the few gallons of gas I use each day will not cause a shortage. On the other hand, if no one else curtails her driving, there is still no reason why I should since I cannot single-handedly avert a shortage. Why would Carter stake political capital on an action of this sort?* Are there really that many people who will do something just because the President asks them to? Such an appeal could work if enough people played a contingent strategy: "I'll lower my driving and see if most others do." If enough people tried this approach, there would be substantial cooperation and the cooperators would all be satisfied that their contingent strategy was working. And indeed it would be working, in the same way that in Graph 6.2 as few as 3 out of the 10 players can switch to C and together cause a net improvement in their own payoffs.

The broadcast nature of communication in this example is crucial. Presumably, in order for each individual to cooperate he must believe he is part of a movement big enough to succeed. Thus it is important that many people start at once. It would be hard for a bandwagon to get started at the level of individual citizens' choices without mass communication.

7.3.2 Structure

The success of unions can be traced in large part to a change of rule, in particular, a rule that involves the notion of "excludability." Let us briefly note how excludability can relate to a "Prisoner's Dilemma" situation. In "Playground" in Section 7.1.1, no one could be excluded from the park. Since nongivers were not excluded, i.e., they received the same benefits as givers, there was a temptation not to

* The Carter energy proposals of April 1977 were of course not restricted to cajolery in the form of appeals to national pride and national priorities. They included such other techniques as economic incentives toward smaller cars, and the threat of future gasoline price hikes if gasoline consumption were not lowered to meet certain targets. Since no single decision-making citizen has more than a minuscule likelihood of determining whether the nation's target is reached, inconveniencing yourself by driving less remains a dominated strategy. However, it is conceivable that if everyone started using buses, service would be expanded, thereby making any inconvenience less noticeable than the savings in dollars.

give. One could say that the nongivers receive a free ride, and in fact such players are often called "free riders." Now suppose that some, but not all, of the workers of a firm belong to a union. They pay dues to hire a negotiator who tries to secure them higher wages. If the negotiator succeds, both members and nonmembers will get the higher wages; the nonmembers cannot be excluded. Staying out of the union is thus a dominant strategy, since the dues (or strike threat) from just one more member will have very small probability of making the difference between success and failure of the attempt to get a raise.

In order for unions to deal successfully with this problem a rule change was needed. They obtained legalization of the closed shop; under this rule, union membership is a requirement for being employed by a firm that has a contract with a union. Thus, since nonmembers cannot be excluded from benefits, they are excluded from the job altogether. Another ploy is to obtain group rates for union members on charter flights and health insurance and exclude nonmembers from these secondary benefits.

Another kind of rule that can be useful is one that allows players to submit contingent decisions. New York City's financial crisis that began in 1976 gave rise to an offering of city bonds in which buyers were allowed to make contingent decisions. When a corporation or city is financially stable it simply offers its bonds for sale on the open market. With New York's financial future in doubt, however, the bonds in this offering might have become worthless unless the entire offering was sold. On the open market, with nearly everyone fearing the worst, there would have been a self-fulfilling prophecy of failure. The city therefore assembled a package deal in which large banks and pension funds were able to buy in simultaneously, with each one's decision contingent on all the others.

Mutually contingent decisions would have been helpful to our park contributors in "Playground" (Section 7.1.1). The technique of using "matching contributions" for charity involves a kind of contingency. Usually, however, it is not practical to make the decisions of many small contributors to charity contingent on each other.

Package deals can be used in other situations where a critical mass needs to be assembled. Recruitment may be difficult for a government agency or academic department with a reputation for being a dull place to work. But if it can somehow commandeer enough resources, it may succeed in hiring several "exciting" people at once by allowing them all to make their decisions contingent on each other. Future recruiting may then become even easier. In terms of X-shaped graphs like Graph 7.8, the action moves over to the right side of the crossover point, where self-interest coincides with group benefit.

Instead of altering the rules—as in the case of the union closed shop or the package deal for bonds—we can try to manipulate another structural feature, payoffs. Two complementary uses of payoffs to promote cooperation are somehow to raise the payoff to cooperators or to lower the payoff to noncooperators.

Governments, especially national ones, have a well-known way of raising and lowering various payoffs, namely the lowering and raising of taxes. This effective and very important way of altering payoffs was treated at some length in Section 6.2 in connection with pollution games. Because those examples were all nonsymmetric, two-person games, they may seem remote from the multi-person, symmetric dilemmas described in this chapter. However, the notion of negative externality, which underlies the pollution games, is closely related to "Prisoner's Dilemma," as the following game shows.

● **Game 7.2 Mutual Pollution**

Two cities border a lake. Each city carts its garbage to a remote dump but is considering dumping it in the lake instead. Let one "utile" equal $10,000 per year of expense or damage. The decision of a city to dump in the lake has two effects: (1) a saving to the dumper of 3 utiles in expense, and (2) a loss of 2 utiles of citizen recreation to each city.

In this game a player can gain one utile (net gain = 3 – 2), thereby imposing a *negative externality* of two utiles on the other player. When the results of both cities' choices are added, the combined outcomes form the "Prisoner's Dilemma" of Matrix 7.2. The separate effects of each city's decision appear around the edge of the matrix of this separable, symmetric game.*

			City B	
			C **(0, 0)**	**D** **(−2, 1)**
City A	**C: cart it away**	**(0, 0)**	0, 0	−2, 1
	D: despoil the lake	**(1, −2)**	1, −2	−1, −1

Matrix 7.2 "Mutual Pollution."

* Of related interest is the 1978 headline "New York and Philadelphia Can Dump on N.J., High Court Rules." A 1973 New Jersey law restricting outsiders from using private N.J. landfills "infringed on interstate commerce," according to the U.S. Supreme Court. (*Los Angeles Times*, June 24, 1978.)

Now suppose the size of the negative externality is less than the net ("internal") gain to the decision-maker. This is the case in "Noisy Neighbors" (Section 4.2) where the negative externality of one unit is outweighed by the decision-maker's gain of three. As a result, that game is not a "Prisoner's Dilemma" but "Convergence." The cities on the lake would also be playing "Convergence" if carting costs were very high or if potential damage to the lake was slight. We saw in Section 6.2 that negative externalities cause net loss to the group as a whole only where marginal damage is greater than marginal cleanup cost. We now see that this same concept of net gain or loss to the group as a whole determines whether the mutual polluters in Game 7.2 are involved in "Prisoner's Dilemma." (For a numerical treatment of this point, see exercise 2 at the end of this chapter.)

The extension to the multi-player case is straightforward, provided that the payoffs are separable (additive). In fact, we have already seen a three-player example in the contribution game "Playground" (Section 7.1), although there the externality is positive rather than negative. The payoffs of that game can be reinterpreted to form a simple mutual-pollution game. Suppose that the same three families again are each deciding whether or not to spend $30 for the community well-being, but now the money is to go for smog-control devices for cars. If one family gets a device (for its own car) air quality in the community will be improved for everybody by an amount equilavent in value to $20 each. Assuming that improvement in air quality can be added up, the appropriate payoff graph is once again as in Graph 7.1c. Thus, the positive externalities of contributing are equivalent to *removal* of the negative externalities of polluting.

In a multi-person game the combined damages to all must be compared with the total cost of avoiding those damages. When damages exceed the cost of avoiding them, we have (in the symmetric case) a multi-person "Prisoner's Dilemma" structure. Moreover, that is just the condition—damages exceeding the cost of avoiding them—that suggests the need for government action (Section 6.2).

The type of government action in this particular case could equally well be a tax or an edict requiring cleanup equipment. But this is only because our example has two simplifying properties: it is symmetric and there is only one means of controlling the externality. But life is not so simple; typically there will be a variety of pollution control devices, some more effective, and also somewhat more costly, than others. Some cars or some industrial plants will be dirtier than others or more suited for particular kinds of devices. As shown in Section 6.2, it is the tax on pollutants that can take all these factors into account, allowing each polluter to use his own information to decrease his pollution to the point where his marginal cost equals the rate of the tax. In this way cleanup will be done most where it is cheapest to achieve substantial improvement.

Governments are not the only ones that can raise and lower payoffs. Labor unions may offer loans to their members and community television stations may offer charter flights to contributors. Large groups can use their size to obtain commercial discounts, thus diminishing the sacrifice of joining them and offering an incentive to cooperation. The other side of the coin, penalties to noncooperators, can be rather annoying when they are introduced simply to stimulate cooperation. For example, many telephone companies are presently waging a cost-cutting campaign to get people to use the telephone directory instead of dialing for information. When you dial the information operator you must first listen to a recording, "If you've checked your directory and are unable to find the number you want, please stay on the line. . . ." Listening to this little sermon is a penalty in time rather than money. Evidently it is working better than the earlier tactic of urging everyone to voluntarily cut back on information calls. Thus, in this case a payoff change was more successful than cajolery.

Payoffs need not be symmetrical, and in reality they probably seldom are. As already noted, contribution games are asymmetric because some people have more money to give. The master of ceremonies in *Radical Chic* (Wolfe, 1970) capitalized on this asymmetry by starting at the top, calling for the largest contributions first. Shortly after the first widespread East Coast blackout, Yankee Stadium announced it was cutting back in its lighting—another case of a "big" player setting an example for smaller ones. During the 1977 California drought it was revealed that one-third of San Clemente's water consumption was going to a golf course, hardly an incentive to the smaller players to take shorter showers and fix leaky toilets.

The technique of variable pricing is useful in handling congestion problems (Section 7.1.2). The telephone company has for decades charged more during business hours than on nights and weekends. The same principle has more recently been introduced on some toll bridges during rush hours. On city streets, where tollbooths are impractical, barriers can be used to extract a payment in time (Berkeley, California, has done this). Many electric utility companies are asking their customers to run their dishwashers at night, to even out the load on generators, but variable pricing could in principle do the job more effectively. However, to use variable pricing, the electric company would need to know *when* you used your electricity, and the electric meters currently in general use do not record the time at which electricity is used, only the total.

Group size can be as important as prices and other payoffs. Common sense and informal observation suggest, and laboratory experiment confirms, that small groups cooperate more readily than large ones. One might therefore expect to facilitate cooperation by making use of small groups where they already exist, breaking down large groups into small subgroups, focusing attention on them as far as possible, and putting up entry barriers to keep small groups from growing larger.

The principle of seeking out the smallest existing organizational unit might help solve the problem of grade inflation. This problem can be approached at the level of the department, the school or division, the campus or the state-wide university system, and indeed it is widely discussed at all levels. However, the dilemma aspects of the situation, noted in the preceding section, suggest that we should look to the departments.

In addition to simply assigning responsibility to the smallest available natural unit, one can publish statistics comparing the performance of similar units. In their campaigns for alumni contributions, some colleges present the previous year's total of gifts by each graduation class. The hope is that giving will be stimulated by class loyalty. This technique could readily be adapted to a variety of other situations. Apartment buildings, for example, could have a monthly account of electricity consumption by floors posted in the lobby. A moment's reflection on the fascination held by baseball standings and fluctuations in stock market prices will show that such an "energy derby" would not likely be ignored by the tenants. Also notice that privacy is not invaded at this level, although there is something of a trade-off between privacy and efficacy of the technique: the larger the group (the floor) the better will the behavior of any individual be concealed, hence the greater the privacy, but hence also the less the incentive to the individual to work toward group success. During the 1977 drought in California a number of cities posted large colorful signs showing targets for water-savings and weekly usage rates.

Entry of new players and barriers to entry into a game are matters of interest to both oligopolies and the Nuclear Club. An oligopoly is a small group of firms that, together, virtually control a market. The Nuclear Club is a small group of nations that, together, virtually control the manufacture of nuclear armaments. In each case the members of the group have a common interest in keeping the group small, in part so that cooperation (tacit or explicit) will be more manageable. We shall discuss the two cases in turn.

An important question in the study of oligopolies is under what conditions it is economically possible for a new firm to enter the market. A closely related question is how and why existing firms keep new firms out. Some of a new firm's customers will be new purchasers of the product, but some customers will be stolen away from existing suppliers. A new firm can woo customers of existing firms through price, advertising, or product variation, all of which can be thought of as the "D" strategy. A new entry in an oligopoly is thus a troublesome event, not only because it increases group size but also because it brings in a relatively desperate player, one who cannot afford to cooperate with the existing oligopolists.

Going into business takes capital, and going into a big business takes a lot of capital. Many people have the necessary courage and ambition, but few have the capital. Capital is therefore a barrier, or at least a hurdle, to entry. Companies in an oligopoly engage in two activities that raise this hurdle considerably: advertising and proliferation of brands. Advertising is of course a heavy expense for existing firms, but it is a much greater burden for a new firm whose limited capital is needed for investment in plant and equipment. Proliferation of brands refers to the fact that a single manufacturer, say of cigarettes, will produce many brands all competing against each other. This policy means an added expense that could be avoided if all the manufacturers of a product refrained from it. Nevertheless, the practice makes it difficult for a new firm to find a competitive variation for a product.*

The entry of new players during a game has not been modeled in the games we have examined, since we have always kept the number of players fixed. Experimental games typically do not allow entry, but in experiments that have compared games of different sizes, cooperation has been shown to be greater the smaller the group (Section 9.5). This suggests that oligopolists might want to discourage new entrants to facilitate joint cooperation in such matters as the tacit agreement not to have price wars.

Communication, which can be highly useful to would-be cooperators in "Prisoner's Dilemma," is a tricky subject for oligopolists, who do not wish to be found colluding in violation of antitrust laws. (The trade journals, however, sometimes publish subtle suggestions about the need for price increases.) In any event, information about the other players' current pricing choices is available daily, while information about payoffs (profits) is revealed in quarterly reports. With communication and information thus quite freely available, one might expect very little competition, i.e., greater cooperation.

---

* Both advertising and brand proliferation have the effect of bringing in new customers to the industry as a whole. The net gain to the industry from this effect may be positive up to a point, but probably is negative in the region where oligopolists are making their decisions.

Can this view of oligopoly provide any insight into the international armaments situation? Here the role of the new entrant is played by the nation about to "go nuclear," either through its own research efforts, by purchasing nuclear arms, or by purchasing nuclear equipment for peaceful purposes and then converting to military uses. The major powers, like the oligopolists, have an interest in keeping the game small by keeping out new players. They can do this by continually "raising the ante." Although many nations now have nuclear arms, the price of entry now also includes delivery systems capable of penetrating very sophisticated defenses. Thus, the arms race "game" between the United States and the Soviet Union has, as a result of uncooperative play, been effectively kept to two players. In both the arms race and oligopolies, then, competition has the side-effect of raising a barrier to entry.

A major national debate of 1968–1969 surrounded the ABM (anti-ballistic missile) system. Insofar as such a weapon would render impotent the delivery systems of all nations other than the U.S. and the USSR, it would have the temporarily stabilizing effect of keeping the arms race game small by raising barriers to entry. Reasoning along these lines, some supporters of the ABM argued that whether or not the ABM afforded a strategic advantage vis-à-vis the Soviet Union, it would at least keep the U.S. invulnerable to China.

The notion of entry barriers links the stability of the U.S.–USSR situation with another important aspect of arms policy, namely selling and giving sophisticated arms to friendly nations that would otherwise not have them. Such gifts and sales have often been criticized for promoting arms races between minor powers. But the entry barrier argument raises a new objection to weapons dispersal to other nations. This is the prospect that the arming of smaller nations by one superpower may be perceived as strategically threatening to the other, who might then feel justified in building ABMs, and this of course would undermine the stability between the two superpowers. (Recall the argument in Section 4.3 that ABMs, though apparently defensive, can create a temptation to a first strike, since the attacker might hope to block retaliation sufficiently.) Thus, if the entry barrier argument is accepted, weapons dispersal to minor powers should be curtailed.

Exercises

○ **1** Consider the following symmetric three-player game. Each player has the option of fertilizing a "money tree" by throwing 5¢ on the ground. After their decisions whether to throw or not have been made, the tree "absorbs" all money thrown to it and creates 6¢ as "fruit" for each 5¢ it gets. All fruit must be divided equally by the players, regardless of who threw fertilizer.

(a) This game is a "Prisoner's Dilemma." Which option is "C"?

(b) On a single set of axes, draw C(n) and D(n).

(c) Now consider a different game, one with four players. Each of them must choose whether or not to contribute a dime, which will result in 24¢ worth of fruit. Draw the graphs.

(d) In either game, suppose that no one is going to contribute, but then two players, acting jointly, change their minds. Will they be better off if they do? (Answer this question for each of the two games.)

○ **2** Three cities share a lake. As in Game 7.2, if one city puts its garbage in the lake, each city loses 2 utiles.

(a) Let the amount a city saves by dumping in the lake be 3 utiles for a net gain of 1 (= 3 − 2). Draw the C(n) and D(n) graphs.

(b) Now suppose the amount saved is 7 utiles for a net gain of 5 (= 7 − 2). Draw the new C(n) and D(n). C still represents "carting" but no longer means "cooperation," since this is not a "Prisoner's Dilemma."

(c) What is the equilibrium outcome in part (b)?

(d) Is that equilibrium deficient? Explain.

(e) For a single city's decision with the potential savings in part (a), compute the total effects to itself and others, that is, the net social benefit (with costs as negative values of benefit).

(f) Same as part (e) but with potential savings as in part (b).

○ **3** Smiley Wiley is running for mayor and his supporters are organizing his one big fund-raiser, a $25-a-plate spaghetti dinner. First they arrange to obtain recycled spaghetti, which costs them nothing. Then they examine the past behavior of the townspeople and deduce that there are exactly 100 potential contributors each of whom will be happier if Wiley wins than if he loses, by an amount of happiness equivalent to $50. If all 100 of these people attend, Wiley will have a campaign war-chest of $2,500 and will surely win. If he gets fewer contributions, his probability of winning will fall off proportionately. Thus, if only 75 people come, his probability of winning is 0.75, etc.

(a) Draw the C and D functions on a payoff graph. Label fully. (You may ignore the distinction between 99 and 100.)

(b) No one wants to miss a good party. But, of course, if no one comes it will not be a good party. In fact, just how good a party it is will be exactly proportional to how many people come. Each person's enjoyment of the party goes up by the equivalent of 50¢ for each additional person present. Draw a new graph of C and D payoff functions.

(c) Does the situation described in part (b) have an aspect equivalent to a "backpatting" factor, as in Section 7.1.2? Briefly explain your answer.

(d) Hope Spendthrift is one of the 100 potential attenders. She is sitting at home deciding whether or not to go. Having drawn her payoff graph she is now trying to guess how many others will go. She will go only if she guesses a number that makes C (going) more attractive than D. About how many people must she think are going in order for her to decide to go?

(e) Ms. Spendthrift, of part (d), concluded that the situation was hopeless; she believed that only a few (3, 4, or 5) people would come. Now, however, she has just learned that there will be a door prize of $500, given out by lottery *only if* 10 or fewer people attend. Draw the new payoff graph, including the additional expected value from this new factor.

(f) What feature of this new payoff graph suggests that hope should spring eternal in Ms. Spendthrift and that she should attend?

○ **4** **(a)** Name some situations in which one or a few players can spoil things for everybody.

(b) Sketch C(n) and D(n) graphs for such situations.

(c) What are ways of keeping people from spoiling things? (Try to generalize.)

(d) Name some situations in which cooperation by one or only a few players can make things nice for everybody.

(e) Sketch C(n) and D(n) graphs for such situations.

(f) What are some incentives we can provide for cooperators in such situations.

○ **5** For the Women's Lib situation depicted in Section 7.2.2, give:

(a) The players.

(b) The alternatives available to each.

(c) The dominant strategy.

(d) Assumptions about preferences that underlie the answer to part (c).

(e) The deficient equilibrium.

(f) Assumptions about preferences that underlie the answer to part (e).

○ **6** **(a)** For Graph 7.8 answer the following:

(i) How many equilibrium points are there (0, 1, 2,...)?

(ii) Describe the location of each equilibrium point on the graph.

(iii) State whether each equilibrium point is deficient or not.

(b) Answer the same questions for Graph 7.6.

○ **7** Costs of bargaining are mentioned in Section 6.2. One of these costs has to do with the fact that there are many sufferers affected by a polluting plant, and these sufferers are involved in a "Prisoner's Dilemma" game among themselves. Explain what this means in nontechnical language that would be convincing to an intelligent student unfamiliar with game theory.

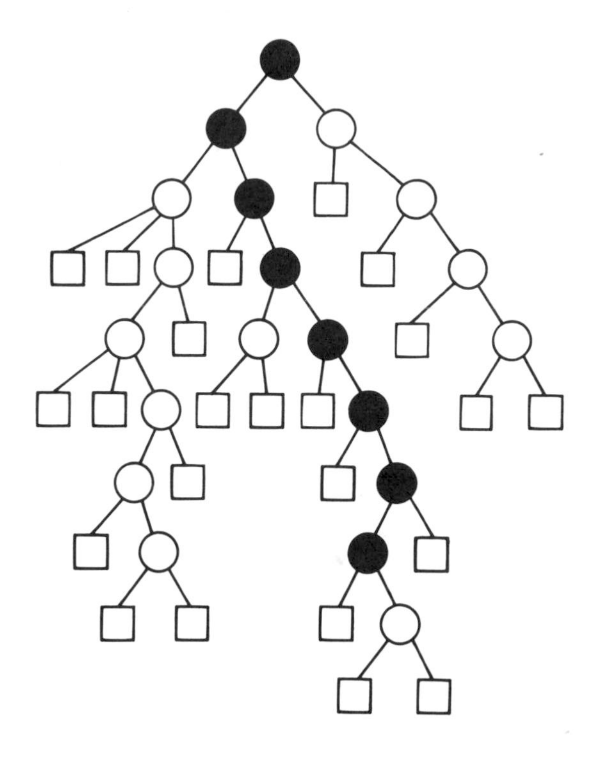

8 Political Games

Coalition-formation and voting are two important political activities that have attracted a lot of game-theoretic analysis. Like the dilemma games of the preceding chapter, both of these involve more than two decision-makers. However, there are major structural differences between dilemma games and political games involving voting and coalitions. These differences in turn give rise to very different analyses.

Excludability is one major area of difference. The dilemmas were the direct result of an inability to exclude noncooperators from the benefits of cooperation. However, when a political coalition forms and is successful in gaining control of government, the remaining individuals or parties most certainly cannot expect to enjoy any benefits. These excluded individuals or parties can, however, seek to undermine the successful coalition; indeed, they are strongly motivated to do so precisely *because* they are otherwise totally excluded from the benefits. Thus the paradoxical aspect of coalitions is their instability. These matters are discussed in Section 8.1.

The situations in this chapter differ in other ways from dilemma games. In dilemma games players simply have to choose whether or not to do some activity, or how much of it to do. Political decisions may be much richer in structure than this. Thus, a voting strategy may involve deciding that you will vote for some amendment and then vote for the main motion only if the amendment passes. In forming a coalition, you may choose among various groups of other players and, having joined a group, must agree with the others how to further slice whatever slice of the pie you can get for yourselves. Voting strategies not only reflect the complexity of parliamentary rules but also are influenced by the availability of coalitions. Section 8.2 deals with the decision structure of voting.

8.1 Coalitions

The formation of coalitions surely has much to do with communication, so it may provide some perspective to review the role of communication up to this point. The potential for communication becomes progressively more complex as one moves from two-person, zero-sum (or constant-sum) to two-person, variable-sum, to multi-person games. In two-person, zero-sum games there is nothing for players to talk about since their interests are diametrically opposed. In variable-sum games two players, with their interests only partially conflicting, can indeed negotiate fruitfully. While they may encounter difficulty in finding particular combinations of demands and concessions that will leave both of them satisfied, they at least have no trouble deciding who to talk to—each must talk to the other. In contrast, in coalition-formation, deciding who to talk to is a key element of the situation.

The kind of game used as a model of coalition-formation situations is called an N-person game in characteristic function form. The characteristic function states how much payoff each possible set of players can get by acting together. In a particular subclass of these games, the so-called simple games, the payoff for each coalition is either 1 or 0, corresponding to winning control of the government or not being able to do so, respectively. The central questions are (1) which coalitions form, and (2) how the members of a coalition divide their joint payoff.

The first subsection describes a much ballyhooed situation in professional sports, the reentry of the free agent. Though this example is economic and not political in nature, it will turn out, in the second subsection, to be equivalent to another situation that is political. This equivalence is expressed through the characteristic function. Of the two central questions mentioned above, the first, that of which coalitions will form, may seem logically prior. On reflection, though, how can you decide who to join up with until you consider how big a share they will agree to give you? Thus, game theorists have seen fit to bypass the first question and ask how the booty might be divided. This topic is approached through notions of stability, asking whether a certain division of wealth would leave some participants unhappy, and whether they could plausibly act upon that unhappiness by setting up a new coalition. Thus, the first question resurfaces.

If a new coalition does form, it may draw upon dissatisfied players from other coalitions. If the coalition to be formed appears to be as unstable as the one being broken up, players may have misgivings and try to look further into the future. Such are the deliberations of Section 8.1.3. The last subsection is specifically political. It surveys theories of political coalition from the game-theoretic perspective developed in the earlier subsections.

8.1.1 The New Baseball Game

A funny thing happened to some baseball owners on the way to the bank in 1977. Suddenly it turned out that they were paying much higher salaries than in 1976. Of course, it was not all that sudden and it did not exactly just "happen" to the owners, in the sense that each salary was negotiated by the owner who paid it. But what did happen to the owners—something that they did not deliberately choose either individually or as a group—was an overthrow of the old rules of doing business. Specifically, the "reserve clause," which bound a player to a single team until he was sold by the owner, was found to be illegal by the courts. What emerged in its place is a system in which a player can become a "free agent" by playing out his contract plus one additional year.

The effect of this rule change was overwhelming. It turned a two-person negotiation between a player and one owner into a multi-person situation. A free agent can bargain with several different owners simultaneously, so that the owners are, in effect, placed in the position of bidding for the services of the free agent. If the owners were to get together and present a united front to a player, then the game would revert to its previous two-party structure; thus, it is significant that owners are legally forbidden to collude.

Perhaps the most important thing to notice about the situation is that the major event occurred in 1976, not in 1977. That is, the game of "Free Agent" was invented, its rules established, in 1976.

Once the rules were in place, the actual playing of the game, the negotiating of specific players' contracts in 1977, was bound to be sharply altered from the play in previous years.

This change in circumstances will be expressed below in simplified form as a change from one particular game to another. The actual change in outcome, the jump in salaries from one year to the next, will turn out to be consistent with the game-theoretic solutions to the respective games. (This is not to say that game theory knows all, just that it gets this one right.) Let us see how the analysis goes.

The old game, "Player-Owner," is a straightforward, two-person, buyer-seller negotiation. Player has a lowest price, L, for which he will sell his services; below that, he prefers to raise soybeans or sell real estate. Owner similarly has a highest price, H, that he is willing to pay; above that he will seek to deal with some other player (presumably a minor league player). Somewhere between H and L, presumably not too far from the halfway point, an agreement can be reached.

In "Free Agent" there is again the player with some lower limit, say $100,000, but now there are any number of owners each with his own upper level. For simplicity, we take the case of two owners each with the same level, $200,000. We assume that the owners are legally forbidden to collude. The player has some leverage because no deal can go through without him, but a deal can go through without one of the owners.

Suppose for a moment that Player and Owner #1 are tentatively considering a contract at the midway point, $150,000. Player can raise an objection to this deal by pointing out that Owner #2, who is now left out in the cold, could easily be enticed by Player into a contract at $175,000 since that would improve Owner #2's lot by $25,000 ($200,000 − $175,000). Owner #1 cannot raise any such objection since he has nowhere to turn, having no gainful way to do business with Owner #2 alone. (In fact, he is legally constrained from talking to Owner #2.) This reasoning, if accepted, leads to the conclusion that Player can object to anything under $200,000 and so could obtain that amount from one owner or the other.

One thing about the "Free Agent" game that probably makes it startling to so many people is that it reverses the usual roles of management and labor. One thinks of a business as typically having a job opening for which a number of potential workers are available. In such a case it will be management that has the strong bargaining position, provided that the applicants for the job cannot communicate with each other. Although the applicants are not forbidden to communicate, they are nevertheless unable to do so because no applicant knows who the others are.

Worker-management games can have at least four combinations of players: one business and many workers, the reverse ("Free Agent"), one business and one worker (simple contract negotiation), and many businesses and many workers. The first three we have already looked at, so let us look at the case of many workers and many businesses. As in "Free Agent," each player has a limit: workers have lower limits on what they will accept and businesses have upper limits on what they will pay. If the numbers of both workers and businesses are very large, then there is what the economist calls perfect competition and the market price for labor will reach equilibrium at the wage w, where the number of workers willing to work for as little as w just equals the number of businesses willing to pay as much as w. A game-theory result similar to this appears in Game 8.8 in Section 8.1.3.

8.1.2 The Characteristic Function

In this section we introduce a political situation and compare it to the economic example of "Free Agent." This is accomplished with the aid of a very important device, the characteristic function.

● **Game 8.1 City Government**

A city council has three members. Whichever of them got the most votes at election time is also the mayor. Their most important decision is the selection of a city manager, who must be approved both by the mayor and by a majority of the council.

In this situation the mayor can join with either of the other council members to pick the city manager. The other two on their own cannot select the city manager, though they can band together to prevent any selection from going through.

The similarity between "City Government" and "Free Agent" is strong. In fact, from the point of view of payoff structure they are identical. One specified player, the mayor or the free agent, can form a mutually rewarding two-person coalition, but the other two players can get nothing from a coalition other than threat potential.

These similarities can be made particularly clear by a new notation for presenting payoffs. This notation, which takes the place of trees or matrices, is known as the characteristic function. It can be regarded as yet another translation, in the sense of Chapter 2, but it is worth noting that a lot is "lost in translation" in going from trees or matrices to the characteristic function. This loss can be useful if the part that is salvaged is the most important or of the greatest interest. What the characteristic function focuses on is the potential benefit of each coalition that might be formed. More precisely, it states how much each coalition could guarantee itself. On the other hand, it omits information about the sequence of moves (as encoded by trees) and about the range of over-all strategy choices (as represented in matrices). A simple example will make these points clearer.

Game 8.2 Three-Way Pile of Four Stones
Player A goes first and takes one or two stones, then player B, then C, then A again if needed. Whoever takes the fourth (last) stone wins.

This game will look familiar from Chapter 2. It is most straightforwardly expressed as a tree and can be converted into a matrix by the methods of Chapter 2. It is a constant-sum game in that there will be exactly one winner. Thus, if the winner's prize is a dollar then payoffs are 1 to the winner and zero to the two losers, for a sum of 1, no matter who wins.

In Tree 8.1 it has been assumed that if a player can win in his current move he will do so; thus, when two stones remain, the player

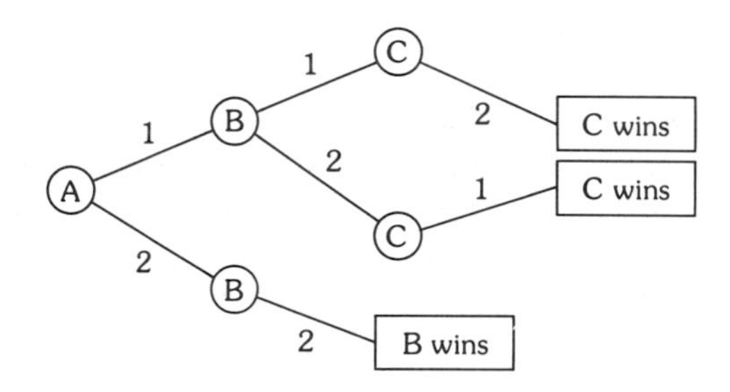

Tree 8.1 "Three-Way Pile of Four Stones."

whose turn it is will take two. With this assumption, it is clear from the tree that C never has a significant choice. Therefore the resulting matrix (Matrix 8.1) reflects decisions only by A and B. The entries in the cells of the matrix are the winners' names. Thus the C in the upper left stands for a payoff of 1 to C and zero to the others.

		B (if A takes 1)	
		1	**2**
A	**1**	C	C
	2	B	B

Matrix 8.1 Translation of Tree 8.1.

The characteristic function form of the game states what each individual or group can insure that it will get. It is clear from the matrix that player A cannot win. However, since A has complete control over whether B or C wins, it follows that neither B nor C acting alone can insure a win. The characteristic function expresses the impotence of individuals acting alone in this game by assigning them each a value of zero: $v(A) = 0, v(B) = 0, v(C) = 0$. Notice that if no coalition is formed, someone will still win; but no one can guarantee a win. The characteristic function reflects what was called a security level in Section 6.1.

A coalition of two players can do better than this. Thus, A and C can decide to work together so that one of them will win the prize, with the understanding that they will divide it in some mutually agreed proportions. They have the power to do this since A can take one stone and then, no matter what B does, C can win. Similarly, A and B can insure victory by forming a coalition and so can B and C. Thus, any coalition of two players can insure a win. (For completeness one also must consider a coalition of all three, which of course would also insure that one of its members wins.) The characteristic function, including the values already given, would thus be

$$v(A) = 0,\ v(B) = 0,\ v(C) = 0$$
$$v(AB) = 1,\ v(AC) = 1,\ v(BC) = 1,\ v(ABC) = 1$$

Game 8.3 Majority Divides a Dollar

Three people, A, B, and C, are given a dollar to divide provided they can agree on how to divide it using majority rule. (If no two can reach agreement, no one gets anything.)

This game would be rather difficult to express as a tree or a matrix but it is ideally suited for characteristic function form. In fact, it happens to have exactly the same characteristic function as the preceding game. For example, player B acting alone cannot guarantee anything, so $v(B) = 0$. On the other hand, since B and C together can form a majority and agree to divide the whole dollar themselves, $v(BC) = 1$.

Returning to the "Free Agent" game, and calling the player and two owners, respectively, P, X, and Y, we have

$$v(P) = 0,\ v(X) = 0,\ v(Y) = 0$$
$$v(PX) = 100{,}000,\ v(PY) = 100{,}000,\ v(XY) = 0$$
$$v(PXY) = 100{,}000$$

Here we have arbitrarily set each individual's self-guaranteed minimum outcome to zero. The player and either owner can get

$100,000 to split in whatever proportions they can agree to. This value is the difference $200,000 minus $100,000, based on the same reasoning as in the buyer-seller game: just as the value of a lotus leaf increases by changing hands, so does a player's services.

The "City Government" game presents a few conceptual problems. Presumably, if two people have the joint power to pick the city manager, then that will somehow be beneficial for each of them. However, there must be some way to share this benefit. One possibility is to allow what is called a *side payment.* In this example one of the coalition members would let the other pick the city manager and would receive some money, the side payment, in return. Although such political use of money is frowned upon, it points up a convenient property of money: it is *divisible* into very small units. Other kinds of payoffs are harder to divide; you can split a dollar with someone but you cannot each appoint half a city manager.

Perhaps the coalition members in "City Government" can divide the payoff by finding a city manager who is somewhere between their positions politically. One may imagine that this position will be closer to each of the coalition members than to the other member of council, but this may be impossible if the left-out member would be the political moderate. Alternatively, perhaps the city manager will be particularly diligent in administering the pet projects of the two people who were directly responsible for selecting her. This diligence would then be the shared resource that the two coalition members would have to divide.

With some such interpretation, and denoting the mayor and the others, respectively, M, A, and B, we have the characteristic function

$$v(\mathrm{M}) = 0, \quad v(\mathrm{A}) = 0, \quad v(\mathrm{B}) = 0$$
$$v(\mathrm{MA}) = 1, \quad v(\mathrm{MB}) = 1, \; v(\mathrm{AB}) = 0$$
$$v(\mathrm{MAB}) = 1$$

Here we again arbitrarily use 0 as the self-guaranteed minimum, the result of being left out of a coalition. We have not said anything about the case where no agreement at all can be reached, but for simplicity we will assume that this result has the same value, 0. The number 1 is arbitrarily chosen as the value to a coalition of being able to pick the city manager.

To render this game identical with "Free Agent," we need merely restate the "Free Agent" game using DMB's in place of dollars as the unit of currency (the DMB or deci-megabuck is worth $100,000). With this change of units the characteristic function becomes

$$v(\mathrm{P}) = 0, \quad v(\mathrm{X}) = 0, \quad v(\mathrm{Y}) = 0$$
$$v(\mathrm{PX}) = 1, \quad v(\mathrm{PY}) = 1, \; v(\mathrm{XY}) = 0$$
$$v(\mathrm{PXY}) = 1$$

A very important type of game is the so-called "simple game," one in which the characteristic function is always 0 or 1. Both "City Government" and "Free Agent" (with adjusted units) are simple games. However, if one of the owners valued the free agent's services more highly than the others, the resulting game would *not* be simple.

In the view of many the most appropriate application of the simple game is the parliamentary system, in which the national administration is formed by a coalition of parties having a majority in the national legislature. This system is used in many countries of Western Europe. A few numerical examples will point up some of the consequences of using a characteristic function.

Game 8.4 Three-Party Legislature

Party A has 40 percent of the seats, B has 35 percent, and C 25 percent. Any coalition with a majority of the seats can form a government.

Notice that any two parties can form a government, even B and C. In this sense they are all on an equal footing. The characteristic function reflects this equality of coalition potential. Assuming that the value of a winning coalition equals one, no matter who comprises it, we have the same characteristic function as in Games 8.2 and 8.3. Therefore, by looking only at the characteristic function, one would get the idea that this is a symmetric situation, with no party better off than any other, despite the uneven distribution of seats. In fact, this same symmetry of the characteristic function would hold no matter what might be the distribution of seats among three parties, provided no single party had an outright majority. The question naturally arises: is it really true that the extra seats give the larger party no advantage in the process of negotiating a coalition? This is a question for which one may seek an answer in real historical examples of coalition-formation in parliamentary democracies.* One may also look at what players do in the controlled setting of the social science laboratory, as in Section 9.6.

Politics may make strange bedfellows, but there are limits. If one assumes that politicians are not just concerned with gaining power but also with what they can accomplish using that power, then parties with widely different ideologies may find it useless or distasteful to form a coalition with each other. Such is the case in the next example.

---

* Some examples appear below in Section 8.1.4, but we cannot do the topic justice here. The interested reader should consult Groennings *et al.*, (1970) and de Swaan (1973).

Game 8.5 Three-Party Spectrum
A legislature has three parties, each having less than half the seats. Parties A and C are ideologically opposite and refuse to form a coalition. Party B is in the center politically and can work with either A or C.

This game is a simple game and in fact has the same characteristic function as "City Government" and "Free Agent." By our previous reasoning, party B here should be very powerful, like the free agent.

In contrast to the examples examined so far, the next political example involves votes on two separate issues. This added complexity allows some interesting considerations to come into play. (The game is taken from a laboratory experiment carried out by Michael Lieserson; see Groennings *et al.*, 1970.)

Game 8.6 Four-Party, Two-Issue Game
Party A controls 30 percent of the votes in a legislature, while B, C, and D have 30, 25, and 15 percent, respectively. Two bills are being considered, bill X and bill Y. The players' sentiments on these bills are reflected by the numbers in the table below, where −10, for example, means that party A is opposed to bill X to such an extent that its passage costs them 10 points worth of satisfaction, whereas its defeat would bring them a gain of 10.

	A(30%)	B(30%)	C(25%)	D(15%)
X	−10	−10	+20	0
Y	−20	+20	0	+30

In this game any two players from among A, B, and C can muster a majority and therefore bring about the passage or defeat of either bill. Player D is superfluous in the sense that any majority coalition that includes him also controls a majority without him. Suppose players B and C get together and agree to pass both bills. This brings B a net gain of 10 and C a gain of 20. Player B might demand some kind of favor in exchange since he is getting less benefit directly from the deal. This might take the form of a promise with respect to future votes. If points are transferable in some such way, then we can say that B and C can gain a total of 30 to split in whatever proportion they can agree on. Considering some other possible coalitions in a similar way yields the following parts of the characteristic function:

$$v(AB) = 20, v(AC) = 30, v(BC) = 30$$
$$v(ABD) = 50, v(ACD) = 20, v(BCD) = 60, v(ABC) = 0$$
$$v(ABCD) = 30$$

At first glance it may seem ironic that the three players with the most votes (A, B, and C) come up with zero all together, whereas any two of them can obtain 20 or 30. Notice that if A and B form a coalition they will surely agree to defeat bill X, which they both oppose. This result is highly damaging to C. Therefore, bringing C into the coalition involves an ideological incompatibility. On the other hand, D, who does not care at all about bill X, can make a mutually beneficial arrangement with A and B since the latter two are evenly split on bill Y, which D cares a great deal about. Thus, A and B would help D pass bill Y. There would, however, have to be some way for D to transfer to A and B a share of the 30 points that accrue to him.

Part of the characteristic function was omitted above; it will now be supplied. Notice that an individual player acting alone, or even a two-player coalition that includes player D, has no control over the voting and so can be forced by the remaining players to come out negative in this game. Player B, for example, will get -30 if the other players vote so as to pass X and defeat Y. The characteristic function reflects this "worst-possible" thinking by assigning a coalition the value it gets if everyone outside that coalition gets together and does whatever will give the coalition the least possible. In our example these values would be

$$v(\text{A}) = -30,\ v(\text{B}) = -30,\ v(\text{C}) = -20,\ v(\text{D}) = -30$$
$$v(\text{AD}) = -20,\ v(\text{BD}) = -60,\ v(\text{CD}) = -50$$

This worst-possible approach is reminiscent of the Shapley criterion for the status quo point, introduced in Section 6.1. It neglects considerations that might show that although a player (in Section 6.1) or a coalition (here) can be forced to take a low payoff, it may not be in anyone's interest to actually make that occur. For example, B and C comprise a majority, so they can control the voting. They could give A and D a total of -20 (as recorded in the characteristic function) by passing X and defeating Y. However, by doing so they would give themselves a total of -10. They would do much better to pass both bills and give themselves a total of $+30$. If they did so, then A and D would get 0, not -20.

Superadditivity is a property of all characteristic functions. This means that players can do at least as well together as they could separately, since they can always just act as they would have separately and then add up their winnings. In the game under discussion, we have, for example,

$$v(\text{ABD}) \geqslant v(\text{AB}) + v(\text{D}) \text{ since } 50 \geqslant 20 + (-30)$$

Thus, since the coalition AB can guarantee +20 and D alone can guarantee −30, we expect that together they can come up with at least 20 + (−30) = −10, just by acting as they did separately. In fact, they can do much better, guaranteeing 50 by adjusting to each other in the larger coalition. When, as in this case, at least some coalitions yield more than the sum of their parts, the game is said to be "essential."

8.1.3 Solutions and Stability

The idea of finding a solution, or at least of asking what a solution might mean, is a tantalizing one for every type of game, and multi-person games in characteristic function form are no exception. Let us begin with a very simple symmetric example, Game 8.3, in which three players are given a dollar to divide in any way that a majority of them agree to.

Using the bargaining principles for a two-person negotiated solution (Section 6.1), one might propose the symmetric division of 1/3 of a dollar each. All three players would be acting together as a single coalition. We represent this payoff distribution by listing the three players' shares (1/3, 1/3, 1/3). The problem that arises with this proposal, and which could not arise in the two-person game, is that a different coalition has the motivation and the power to break it up. For example, player A and player B can get together and agree to share the dollar equally between themselves, excluding player C. This new proposal is denoted (1/2, 1/2, 0). Players A and B are called the *effective set* for this change because the change gives each of them an improved payoff (thus they have the motivation) and because they are a majority (thus they have the power according to the rules of this game). The new payoff distribution is said to *dominate* the old one. (Note that this is a different meaning of "dominate" than that used earlier. Our earlier use of the word refers to a relation between strategies whereas here we are referring to a relation between payoff distributions among players.)

Now of course this new arrangement is not going to please C, and moreover there is something he can do about it. He can, for example, offer B a better deal, say 3/4 of a dollar, than B is getting from A. That is, C proposes (0, 3/4, 1/4). This payoff distribution dominates (1/2, 1/2, 0), with B and C as the effective set. If we could find a payoff distribution that was not dominated by any others, then that would certainly be a strong candidate for the title of "solution." Unfortunately, no such distribution exists in this game. (A payoff distribution that is not dominated is said to be a member of the core. The reader may wish to prove that there is no distribution belonging to the core of this game, that is, this game has an empty core.)

Failing to find an undominated distribution of payoffs for many games, von Neumann and Morgenstern (1944), the founders of game theory, defined a weaker concept of solution, namely the *solution set.* Rather than a single distribution, they accepted a set of distributions. Consider, for example, the three distributions (1/2, 1/2, 0), (1/2, 0, 1/2), and (0, 1/2, 1/2). A set of distributions is said to constitute a solution in that (1) no member of the set is dominated by any other, and (2) every distribution outside the set is dominated by some member inside the set. We shall verify below that these two conditions are met by our set of three distributions.

The idea of the solution set is that it reflects certain principles that a society may have. The particular set just described might be said to embody the principles that (for this game) two players should split the reward equally. Criterion 2 insures that, starting from any distribution whatsoever, it will be reasonable to expect movement to some member of the solution set. Criterion 1 insures that if players conform to the "principles of society" then they will not keep moving from one distribution to another.

To verify that our set of distributions is a solution set, we check the two conditions. Condition 1 is met because any switch among the three given distributions would benefit just one player. Condition 2 is met because no other distribution (outside this solution set) can give 1/2 to more than one player (since there is only one dollar available). In other words, in every distribution outside the solution set there will be two players each getting less than 1/2, and these two will form the effective set for shifting to a member of the given solution set. For example, (0, 3/4, 1/4) is dominated by (1/2, 0, 1/2), with A and C forming the effective set.

The idea of dominance has played a key role in the analysis so far. It has been suggested, at least tacitly, that if a payoff distribution is dominated then it will be immediately forsaken in favor of another distribution that dominates it, unless the dominating distribution somehow violates some norms or societal principles. Perhaps, though, an effective set of players will not automatically break up a coalition. If the new coalition they are considering forming is likely to be broken up, leaving them out in the cold, then perhaps they will be reluctant to break up the existing situation.

Considerations of this sort are embodied in Vickrey's (1959) notion of *self-policing patterns.* A pattern is a special case of a von Neumann–Morgenstern solution, so it is a set of payoff distributions. Again it is assumed that certain societal pressures favor the distributions inside the pattern, although, again, members of the pattern may be dominated by distributions outside the pattern.

Imagine that some coalition has formed and has agreed to a payoff distribution in the pattern. Now suppose a group of players, with the power and motivation to do so, brings about a new payoff distribution that is not in the pattern and that dominates the original distribution. This new distribution is called a "heresy" and the effective set of players who bring it about is called the "heretical set." What will be the fate of these heretics? Sooner or later, societal pressures can return the game to some distribution that both dominates the heresy and belongs to the pattern. Let us call these distributions the "possible evolutions" of the heresy. It is Vickrey's point that stability of a set of patterns hinges on whether heretics may come to grief in the possible evolutions of heresies. In particular, if there is one heretic who receives less in every possible evolution than he did in the original distribution, then the heresy is "suicidal" for that player. A very stable kind of pattern is one whose member distributions are dominated only by heresies that are suicidal for someone in the heretical set.

A third theory of multi-person games in characteristic function form was developed by Aumann and Maschler (1964). This theory focuses on how a coalition will distribute its payoff if it should happen to form. An example will make the main ideas clear.

Game 8.7 Two-Ticket Game

A couple advertises that they want to buy two tickets to a sold-out concert, and three people answer the ad. A has the best seat, C the worst, while B's is midway between in quality but located at the other side of the hall. The couple decide that they are willing to pay $20 to A and B, or $18 to A and C, or $14 to B and C. The three players are A, B, and C; each of them is unable to attend the concert, so to get more than a zero a player must join in a two-way coalition.

The characteristic function for this game can be taken directly from the verbal description.

$$v(A) = 0, \quad v(B) = 0, \quad v(C) = 0$$
$$v(AB) = 20, v(AC) = 18, v(BC) = 14, v(ABC) = 20$$

Suppose that B begins the discussion by suggesting that he and A make a deal, since they have most to gain; B further proposes an even split of the proceeds, $10 each. In response, A proposes a 15–5 split, pointing out that he is in the strongest position and that he could get $15 by offering a 15–3 split to C, who presently seems likely to get nothing. This of course is a rather limited view of C's role. Once C is brought into the picture, both A and B will be able to make offers to C, in effect bidding against each other. Aumann and Maschler require that A's and B's intended proposals to C be balanced with respect to each other, in a sense to be clarified shortly.

The 15–5 proposal is not balanced because, while it allows A to offer 3 to C, it allows B to offer much more. Specifically, B has a valid objection to receiving only 5 because he could offer C up to 9 in a partnership and still have at least 5 for himself (since B and C can get 14). For A to raise a counter-objection, A would have to match the offer of 9 to C and yet retain 15 for himself. This A cannot do.

The Aumann–Maschler theory calls for a 12–8 split for A and B. We shall see in a moment how these numbers are arrived at. For now, let us suppose that B just happened to propose these two numbers. An objection by A would consist of showing that he, A, could do better elsewhere. He could point out that C would be happy to join a coalition with him giving A more than the proposed 12, that is, C would be offered up to 6 (= 18 − 12). But B can counter-object, pointing out that he too has a possible coalition with C in which he, B, could get more than 8. Like A, B would then be able to offer C up to 6 (= 14 − 8). The counter-objection (by B) is thus exactly as attractive to the third party (C) as is the original objection (by A). Therefore 12–8 is an appropriate division.

To find numbers like 12 and 8 that work out to be a balanced proposal in this sense, let x be proposed for A and $20 - x$ for B (these two amounts add up to 20, which is what A and B can get acting together). An objection by A would consist of pointing out that he can get more than x by offering C any amount up to (but not equal to) $18 - x$. For B the counter-proposal to C could go up to $14 - (20 - x)$. Objection and counter-objection are in balance when

$$18 - x = 14 - (20 - x)$$
$$24 = 2x$$
$$x = 12 \qquad 20 - x = 8$$

The Aumann–Maschler theory does not state which coalition will form but it does suggest how much each player will get in whatever coalition does form. By reasoning similar to that above, if A and C get together they should arrange a 12–6 split, while B and C would divide their gains 8–6. Notice that it turns out that a player gets the same amount in either coalition available to him, so the main objective for a player seems to be not to get left out.

Interestingly, the three payoff distributions endorsed by Aumann-Maschler here, taken together, constitute a solution set in the sense of von Neumann and Morgenstern, as we now show. (This is always the case for three-person games but not always for larger games.) The distributions are (12, 8, 0), (12, 0, 6), and (0, 8, 6). None of these distributions is better than another for more than one player, so none dominates another. Also, any distribution outside the set is dominated by some member of the set. For example, suppose that A and B form a coalition so that according to the characteristic function the sum of their payoffs is 20. Any distribution in which A gets less than 12 is dominated by (12, 0, 6), while a distribution in which A gets more than 12 will leave B with less than 8 and so be dominated by (0, 8, 6). Similar arguments apply if the coalition of A and C or B and C forms. Therefore, the three Aumann–Maschler distributions satisfy the two requirements for a von Neumann–Morgenstern solution set.

Game 8.8 Two Lotus Leaves

Agatha and Semantha each have a petrified lotus leaf to sell for anything over $100. Bertram and Tristram each wish to buy a leaf for anything under $200.

As in Section 5.3, a buyer and a seller can increase their joint well-being by $100. The key parts of the characteristic function are therefore as follows:

$$v(\text{AS}) = 0,\ v(\text{BT}) = 0$$
$$v(\text{AB}) = 100,\ v(\text{AT}) = 100,\ v(\text{SB}) = 100,\ v(\text{ST}) = 100$$
$$v(\text{ASBT}) = 200$$

Suppose that Agatha and Bertram agree to a 30–70 split while Semantha and Tristram arrange a 60–40 division. Agatha can now raise the objection that she can get more than her current 30 by going to Tristram and offering him something less than 70 but more than his current 40. Agatha's coalition partner, Bertram, is unable to raise a counter-objection. He certainly cannot go to fellow-buyer Tristram, nor can he go to Semantha since to maintain his current 70 and entice her away from her current 60 is impossible within the $100 constraint of their potential coalition. Therefore, the current distribution of (30, 70, 60, 40) for Agatha, Bertram, Semantha, and Tristram does not meet the Aumann–Maschler requirements.

Payoff distributions that do meet the Aumann–Maschler conditions here will be all those in which both buyers get the same payoff and both sellers get the same (this result is left for the reader to verify by generalizing the argument in the preceding paragraph). The buyers may get more or less than the sellers. Moving from the world of the characteristic function back to the world of lotus leaves, i.e., from the model world back to the real world, we find that the solution prescribes that a going price, or market price, be established. That price can be anywhere in the range where both buyers and sellers remain interested. The sellers may be better bargainers than the buyers, or vice versa. However, it cannot be the case that one seller is a strong bargainer and the other a weak one. What would then happen, according to the theory, is that the two buyers would, through offers and counter-offers, bid up the weak seller's price until the two sellers had equal prices. This result conforms to the economist's notion of a "market price." If the sellers get together to form a monopoly, they will get the kind of advantage that the free agent has. If the buyers join to form a consumer's cooperative, then they will gain such an advantage. Finally, if both sides are fully organized then there will be two-party negotiation, as with management and labor.

8.1.4 Political Coalitions

The theory of political coalitions as advanced by Riker (1962) uses a characteristic function that accords a payoff of 1 to any coalition that has a majority of seats and 0 to all others. Comparable theories, by Caplow (1956) and Gamson (1961), have appeared in the sociological literature.

The fundamental point of Riker's theory is that if a coalition is big enough to win, it has no need to take in additional members. Indeed, it should avoid doing so since new entries would demand a share of the rewards. Accordingly, if players seek to maximize their individual payoffs, the winning coalition that actually forms should be a "minimal-winning coalition." That is, it should be so small that if any member were removed, the remaining coalition could no longer win.

● **Game 8.9 Four-Party Legislature**

Party A has 40 percent of the seats in the national legislature, B has 25 percent, C has 20, and D has 15. Any coalition with a majority of the seats wins control of the government.

The minimal-winning coalitions in Game 8.9 are AB, AC, AD, and BCD, as the reader can verify. According to Riker, any of these may form, but no other winning coalition will form. Gamson's theory provides a way of choosing among these alternatives. It assumes that players will demand shares of payoff proportional to the resources they provide. Thus, party A can expect to get more reward if it joins with D than with B or C. From a slightly different angle, it can expect B or C to be more demanding than D. Therefore, A prefers D as coalition partner. Similarly, D prefers A to its only other alternative, a three-way coalition with both B and C. According to Gamson's theory, the coalition with least resources will form, in this case AD with 55 percent of the seats. Since this theory makes a more exact prediction than Riker's, it is preferable, provided it turns out to be correct.

However, the characteristic function does not express the notion of least resources, while it does tell us which are the minimal-winning coalitions. We find, for Game 8.9,

$$v(\mathrm{AB}) = 1,\ v(\mathrm{AC}) = 1,\ v(\mathrm{AD}) = 1,\ v(\mathrm{BCD}) = 1$$

These are the minimal-winning coalitions; all coalitions containing these also have the value 1, while all others have value 0. A von Neumann– Morgenstern solution set for this game is (2/3, 1/3, 0, 0), (2/3, 0, 1/3, 0), (2/3, 0, 0, 1/3), (0, 1/3, 1/3, 1/3). This solution set calls for any minimum-winning coalition to form and for A to receive a double share if it is in the coalition. The differences in percentage of seats among B, C, and D are ignored.

Experiments with coalition games (described in Section 9.6) have shown the least-resources theory to be fallible and Gamson has suggested some explanations. One of these has to do with the costs of bargaining (Section 6.2). A coalition of many weak players presumably requires more negotiation to assemble than one with a few strong players. Although no explicit costs are involved, the perceived likelihood of reaching agreement might be affected. The second of Gamson's explanations involves people's beliefs more directly. In Chapter 7, for example, we noted that if everyone believes there will be a run on the bank then indeed there will be one. If the least-resources principle were correct and sufficiently obvious so that people acted according to it, the players outside that coalition would realize that by doing the "obvious" they would naturally lose out. Realizing this, they would stop doing the obvious and try to prevent the winning coalition with smallest resources from forming.

Game 8.10 Variant of Four-Party Legislature
Same as Game 8.9 but with percentages 45, 20, 20, and 15 for A, B, C, and D, respectively.

The two effects just described would have a good chance of materializing in Game 8.10, where the winning coalition with smallest resources is BCD with 55 percent of the seats. If party A figures that smallest-resources coalitions will form, then it will expect to be left out. Realizing this, it acts to overcome this disadvantage. Now the other effect comes into play: while B, C, and D are in the process of getting themselves together and reaching agreement, A can waylay any one of them and strike some bargain.

If this latter effect were sufficiently strong, one might wish to propose yet a third theory, a fewest-player principle. Alternatively, we can seek a resolution among these competing theories by stating something like the following: Some minimal-winning coalition will form; among these, the most likely are those with the fewest players and smallest resources, according to some combined measure.

Two factors that are particularly salient in forming political coalitions are ideology and information. Consider ideology first. We have already noted that parties with opposing or partially opposing positions will not gain much from a coalition if they are more concerned with enacting their programs than with merely being in power (this is the basis for Game 8.5). To account for this effect, a "minimum range" theory has been proposed. According to this theory, parties are assumed to be laid out on a political spectrum from "left" to "right," and those closest together find it easiest to get along. The theory therefore predicts that the coalition that forms will cover a continuous range of the spectrum, with no "gaps."

A left-right spectrum may be inadequate for predicting coalitions in Game 8.6, as Lieserson, who constructed the game, points out. Notice that C and D in that game have quite different beliefs but can easily do business with each other because each one is indifferent to one of the bills (rating it 0) and so each is quite willing to trade votes for the bill it cares about. In contrast, A and B agree on one bill (−10 each) but cannot readily agree on what to do about the other one. In fact, it is impossible to place these players on a left-right spectrum for bill X without misordering them with respect to bill Y.

A lack of party discipline may lead to formation of coalitions considerally larger than 50 percent. In the above examples a coalition wins if its combined total of seats is more than 50 percent. Nevertheless, individual legislators may bolt the party on a particular issue or on a vote of confidence, thereby subverting the power of the majority coalition or (in some countries) bringing down the government. For this reason a bare majority, formed on the principle of "smallest resources," may not be a reliable or stable basis for governing. Any small party within a bare majority coalition would have considerable threat potential, so that even if it were only mildly opposed to majority policy it might be tempted to demand concessions. Therefore unreliability or just uncertainty may lead to overkill in coalition size.

The principal application of coalition theory has been, as noted, to the process of forming a governing coalition in countries with multi-party systems where no single party has a majority. The rewards to be divided are typically regarded as being the cabinet posts and, of course, the office of prime minister. It is not obvious what relative values to attach to each of these positions and, in fact, parties may attach differing importance to the various ministries. Nevertheless, it seems more promising to look at inputs (distribution of seats) and outputs (distribution of cabinet posts) than at the complicated process of negotiation that comes in between. This is the conclusion reached by de Swaan (in Groennings *et al.*, 1970, p. 425) after commenting on the difficulty of studying actual negotiations.

> Little is known about the actual proceedings during the negotiations that lead to the creation of a cabinet coalition. Politicians are not eager to give a full public account of what goes on during the coalition bargaining, since these facts are, almost by definition, embarrassing to them: they involve the gradual adjustment of their electoral promises to those of their prospective partners in order to hammer out a government program that is mutually acceptable and consistent enough to form a basis for actual policy. This is a common dilemma in political life, but in multiparty systems it is the more painful, in that each party had to stress its unique and discriminating features during the preceding electoral campaign in order to justify its separate existence. The secretiveness that surrounds the negotiations and the speculations and self-justifying statements that accompany every new development greatly hamper empirical field work. The data that are supplied by expert observers, historians, and autobiographers, on the other hand, show a multitude of issues, an almost inextricable interplay between accidental circumstances and personal ambitions, principles and sympathies, so as to complicate analysis beyond what is feasible.

Studies of actual political coalition-formation have been carried out for countries as different and remote from each other as Japan, Brazil, Norway, and Uganda (see Groennings *et al.*, 1970). These studies have also focused on a considerable variety of levels of political organization, including parties, tribes, factions, and whole countries. Not surprisingly, no firm or universally applicable conclusions can be drawn. Nevertheless, the kinds of theoretical perspectives mentioned here have helped provide a clearer view of particular cases as well as a basis of comparison among them.

8.2 Voting

Many types of decision structures are only vaguely defined. Voting, however, is a clearly defined decision structure: you vote for this candidate or that one, or for or against a particular bill. The rules about outcomes—majority rule, three-quarters rule, tie-breaking rules, veto-override rules—are also clearly defined. Of course, just because voting choices are well defined does not mean that they are simple. As a voter I may have opinions on a few key issues, but there may be no single candidate who takes my position on all of them. Or I may not be able to obtain sufficient information on an issue to form a clear opinion. As a legislator, although I wish to benefit my constituents, I may not be able to tell what the specific effects of a bill will be. Moreover, it has often been observed that the real decisions in Congress and other legislatures are made long before voting time—in appointments to committees and procedural rulings. Some of these complications turn out to have a reasonably well-defined decision structure of their own that can be incorporated into a game framework. Others can either be regarded as challenges for future study or as warnings of the limitations of game analysis.

Even if voters have full information and clear-cut preferences, some rather sensible-sounding propositions turn out to be false. For example, one might imagine that if each *individual's* preferences were simply ordered, then the *group's* preference, as determined by majority rule, would be simply ordered. In fact, as shown in Section 8.2.1, the group may turn out to have intransitive preferences despite transitivity for each member.

One of the complications mentioned above that can be incorporated into a game framework is voting procedure. One can ask whether and how a change in procedure will affect the success of legislation. Changes in procedure mean changes in the game tree, and these in turn mean that changes in strategy may be necessary or desirable. We will find that some strategies seem more "opportunistic" than others but that a legislator's actual motivation cannot be determined just by looking at her votes. Deliberations of this sort comprise Section 8.2.2.

8.2.1 A Paradox for Two-Dimensional Politics

The simplest imaginable voting procedure is a decision between two alternatives in which a simple majority wins. This case is so simple that there is little to say: the voting is conducted and the alternative with a majority wins. (In the following discussion we will always assume majority rule and an odd number of voters, all voting.) One need not go very far beyond this simplest case, however, to find a situation about which a great deal has been said, namely "Arrow's paradox."*

The simplest form of the paradox can occur when three voters (or voting blocs, or committee members) are faced with three alter-

* Arrow's theorem shows that several apparently sensible criteria cannot all be satisfied simultaneously by any voting scheme (Luce and Raiffa, 1957).

natives, A, B, and C. We assume that each voter has preferences that can be ordered, i.e., we can use an ordinal scale for utilities. Specifically, suppose voter #1 prefers A to B to C and the other voters are as shown in Figure 8.1. Notice that if alternatives A and B are

Voter #1	Voter #2	Voter #3
A	B	C
B	C	A
C	A	B

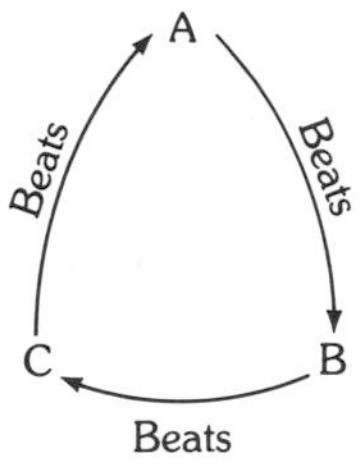

A
Beats
Beats
C
B
Beats

Figure 8.1 A voting cycle. Although each voter has a simple order of preferences, the group does not.

pitted against each other, then voters #1 and #3 will vote for A, so A beats B. Similarly, you can check that B beats C. Now if A beats B and B in turn beats C, one might think that A would beat C. Such would be the case if the *group* (the three voters) had a transitive preference order (see Section 3.3). Paradoxically, however, A loses to C, with voters #2 and #3 forming the majority. Thus, *group* preferences form a cycle in which A beats B beats C beats A beats....

This cyclical outcome is paradoxical in two senses. First, it is surprising to anyone who would guess, offhand, that transitivity among individual choices would yield transitivity among group choices. Second, even if no one is surprised, the three voters are still left with the problem of making a decision (picking an alternative), and there is no obvious way to do this.

Once again, as in the group dilemma games of Chapter 7, the problem of combining individual preferences into a group preference has produced a perplexing situation, though this voting paradox is structurally quite distinct from the multi-player "Prisoner's Dilemma." If one thought that the way out of the "Prisoner's Dilemma" would be found in voting on governmental action, then this voting paradox might be a sobering discovery.

The individual preference orderings in Figure 8.1 are not the only possible ways to order three candidates or alternatives. In addition to the three already given, there are CBA, BAC, and ACB. Now we may ask whether the three orderings that yield the cycle in Figure 8.1 are particularly likely (or unlikely) to occur together. To answer such a question we must make some assumptions about either the voters or the choices they must make (candidates or issues). In fact, we shall make an assumption involving both.

Suppose that all the voters and candidates are regarded as points located at various positions along a straight line and that a voter's ranking of candidates is determined by how close they are to him. Such a picture is conveyed by the idea of a political spectrum from "left" to "right" or from radical to reactionary. A one-dimensional model of this sort often implicitly or explicitly underlies comments by politicians and journalists.

If A, B, and C are laid out as shown in Figure 8.2, a voter to the left of all candidates will find that A is closest to his political beliefs,

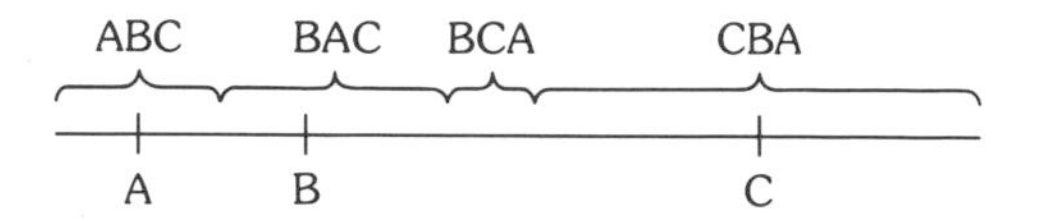

Figure 8.2 A political spectrum with the four possible preference orders. No voter ranks candidate B last.

while C is furthest. Such a voter will therefore prefer the candidates in the order ABC. This same order of preference will be shared by voters slightly to the right of A, who are still closer to A than to B. As shown in Figure 8.2, the preference order ABC will be shared by all voters leftward of the midpoint between A and B.

The next region, containing voters whose preference ordering is BAC, extends from the midpoint between A and B to the midpoint of A and C. The remaining two regions (BCA and CBA) in the figure are determined similarly. Notice that only four of the six possible orderings occur in this one-dimensional model. Nowhere does there appear a voter on this continuum who prefers the candidates in the order ACB or CAB. The reason is that the centrist candidate, B, cannot come last on any voter's list. This in turn is because to rank B behind A one must be to the *left* of B (in fact, left of the midpoint of A and B), while to rank B behind C one must be to the *right* of B, and one cannot be both to the left and to the right of B.

If there really is a one-dimensional political spectrum of this sort, then voting rules can have a big effect on whether a centrist candidate is viable. For example, in the 1970 New York senatorial race there were three major candidates: James Buckley, the Conservative party candidate, Charles Goodell, a liberal Republican, and Richard Oettinger, the Democrat. Buckley won a plurality and thereby was elected. Goodell, the centrist, came in last. In the closing days of the campaign the opinion was widely expressed that Goodell should withdraw in favor of Oettinger, to whom he was closer ideologically. One could argue, however, that the one to withdraw should have been Oettinger, even though he was in second place, since all his supporters would go to Goodell, whereas some of Goodell's might have Buckley as second choice. The opinion that Buckley would have lost a two-way race to a centrist gained credence six years later when he lost his 1976 bid for reelection to a single opponent. His opponent, Daniel Moynihan, was relatively centrist, having defeated the more liberal Democrats Bella Abzug and Ramsey Clark in the primary.

Returning to the question of whether the configuration of Figure 8.1 is particularly likely to occur, we now can say that with our one-dimensional model such voting cycles are not only unlikely but in fact impossible. To see why, consider the one-dimensional situation of Figure 8.2, with its four possible preference orders. Suppose that one of these four orderings were to have a majority of the voters. Then this single majority bloc would determine every possible vote between pairs and their transitive (noncycling) preferences would determine the overall results of voting. Now suppose that no single ordering has a majority. This in turn would require that B beat A, since only one ordering has A ahead of B; similarly, B would have to beat C. But then B would be a clear winner, so there would be no cycle. Thus, there is no cycle either way, i.e., whether there is a majority bloc or not.

The voting paradox is thus ruled out in the one-dimensional case. Can it occur in more complex situations? The answer is yes, as the two-dimensional diagram of Figure 8.3a shows. The entire region labeled ABC contains positions for which A is closest and C is furthest. If we continue to assume that closeness determines preference,

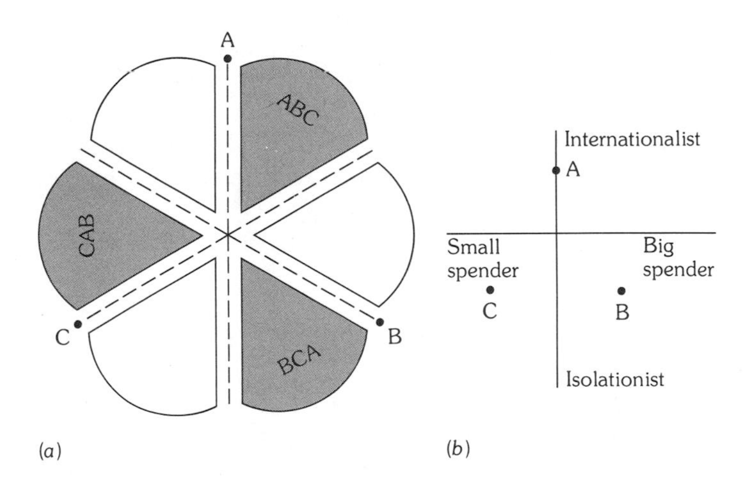

Figure 8.3 A two-dimensional model of possible political positions. In Matrix (*a*), if the three shaded areas each contain one-third of the voters there will be a voting cycle. Matrix (*b*) is a representation of candidate positions in terms of real issues.

then any voter in this region will prefer the ordering ABC. If the three shaded regions contain all voters and no single one has a majority, then the voting paradox is present in this population. In fact, the distribution of voters need not be so uneven. If each of the six "wedges" in Figure 8.3a contained *about* the same number of voters but the three shaded ones each had only slightly more than the three others, the paradox would still be present.

Politics may indeed be two-dimensional or multi-dimensional. An individual may be internationalist or isolationist in foreign affairs, independent of whether he is a fiscal conservative or liberal in domestic matters. Interpreted in this way, the candidates in Figure 8.3a might be as shown in Figure 8.3b, with A being an ardent internationalist and at the same time a moderate spender. Despite this logical distinction we note that the terms "liberal" and "conservative" are applied in both foreign and domestic politics. To the extent that one dimension serves as an indicator of a second, the second dimension will be superfluous. Thus, if all or virtually all internationalists were big spenders and all isolationists were small spenders, and if moreover a person's degree of internationalism always matched his degree of spending, then it would indeed be informative to speak simply of how liberal or conservative someone is.

Even if individual bills were usually one-dimensional, there could still arise a multi-dimensional situation at elections, when a candidate runs on a platform with positions on many bills. In the next example, not only does an "Arrow paradox" of cyclical preferences show up, but in addition we find what might be called a "platform paradox," that is, a winning platform can be formed by combining two losing positions (see Downs, 1957).

Suppose there are three voters, 1, 2 and 3, and two bills, A and B. Voter #1 supports A but opposes B, while voter #2 holds the opposite view. Moreover, each feels more strongly about the bill he supports than about the bill he opposes. Finally, voter #3 opposes both bills, one more strongly than the other. In this situation each bill considered separately will fail since each has only one supporter. Nevertheless, in comparison to a platform opposing both bills, a platform supporting both bills will appeal to voters #1 and #2, since for each of them the support of "his" bill is, by our above assumption, more important to him than opposition to the other bill. (This platform would not, however, succeed in a two-way election against a platform opposing *just* the bill that voter #3 opposes more strongly. This latter platform would, in turn, lose to one opposing both bills, thus completing a cycle.) As with the earlier example, the three voters can be replaced by three groups, none of which has a majority. It is also readily possible to construct examples with larger numbers of groups.

The key to the possible success of a platform of minority positions lies in the importance of those positions to the people who hold them. We might refer to such a platform as a "coalition of passionate minorities."

8.2.2 Procedures, Principles, and the Powell Amendment

A real situation in which the paradox may have been present occurred in 1956 when the U.S. House of Representatives considered a bill for school construction and an amendment to it, the Powell Amendment, denying aid to states with segregated schools (see Riker, 1965). This amendment made the situation distinctly two-dimensional by tying the segregation issue to the question of federal aid to education. What in fact happened was that the Powell Amendment passed but then the entire bill, as amended, went down to defeat.

A quite reasonable analysis of this situation is that there were three voting blocs with the preference orderings shown in Table 8.1,

Table 8.1 Preference patterns for the Powell Amendment.

Northern Democrats	Southern Democrats	Republicans
A	B	C
B	C	A
C	A	B

LEGEND: A = amended bill; B = bill (without amendment); C = current situation (no bill passes).

no single one of them having a majority. Although this pattern of preferences is cyclical, there is in the actual situation one particular outcome that prevails. This is because the *procedural rules* do not treat the various alternatives symmetrically. The usual procedure, as shown in Figure 8.4, is that the amendment is voted on and then the bill itself (as amended, if the amendment passed) is considered. If this final vote on the bill, with or without the amendment, is negative, then the final outcome is C, the current situation. Thus C is still in the running on the final vote no matter what happens on the first vote. In contrast, A, the amended bill, will be a contender in the final vote only if the amendment has passed.

So far we have spoken of preference orders and procedural rules. A third factor in determining the outcome of voting is the

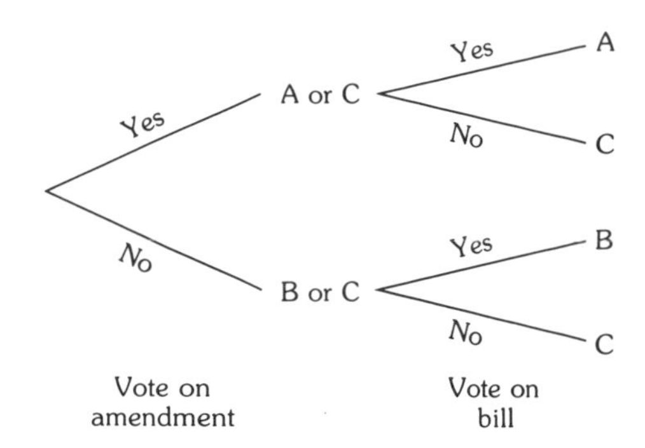

Figure 8.4 Voting procedure when an amendment is proposed (A = *A*mended bill; B = the original *B*ill; C = the *C*urrent state of affairs).

choice of decision principles. The first two examples of decision principles come from general game-theory ideas discussed in Chapter 3, specifically Rationality Principles 1 and 2.

Voting Principle 1
Never use a strategy that is dominated by some other strategy.

Voting Principle 2
Assume that all players will use the same principles.

Taken together, these two principles enable us to work backward through the procedure tree of Figure 8.4, considering the final vote first. If the amendment passes, voters must choose between A and C. The victory will go to C, with Republicans and Southern Democrats forming the majority. For anyone in these two groups to vote for A at this point would violate Voting Principle 1. For these voters, any strategy that calls for voting for A over C at this point is going to be dominated by one that is identical everywhere else but calls for voting for C rather than A at this point. Similarly, we can eliminate any strategy that calls for any player to vote against his simple preference on the second vote.

In this way, Voting Principle 1 completely determines the second vote. Just as C will defeat A if the amendment passes, so in the same way we find that the unamended bill, B, defeats C if the amendment fails. If players know each other's preferences, and if Voting Principle 2 holds, then everyone knows just what will happen on the second vote. To vote for the amendment, then, is really to support C, since everyone knows that, should the amendment pass, C would be the final result. Similarly, to vote against the amendment is to support B. The initial vote thus boils down, in reality, to a choice between B and C. Since two of the three voting groups prefer B to C, we conclude that B will win.

According to this reasoning, the Northern Democrats are supposed to vote against the amendment, even though they like it, on the grounds that it is not politically viable, not really an achievable possibility. They must defeat it in order to salvage the original bill; otherwise they will lose everything. Of course this may sound terribly cynical. One could imagine a more idealistic approach:

Voting Principle 3
Vote for an amendment in the event that you prefer passage of the amended bill to passage of the unamended bill. Otherwise vote against it.

For the Northern Democrats this last principle is incompatible with the more calculating approach described earlier. However, for the Southern Democrats the two approaches yield the same decisions. The same goes for the Republicans, who certainly must have calculated well since they got their most preferred outcome. On the other hand, note that their votes are consistent with the idealistic Voting Principle 3.

It has been suggested that in the actual event the Republicans' true preference order was not that given in Table 8.1 but that in fact they preferred C to B to A. Such a reordering would not be in conflict with the first two principles but would make those calculations incompatible with Voting Principle 3. This is a rather damaging allegation, that they not only held a racist preference order but in addition voted insincerely. However, their votes alone do not tell us what decision principle they used unless we have some independent method of finding out their true preference order. To put it the other way around, their votes do not tell us what they really preferred, unless we can find out what their decision principles were.

This situation is ripe for the formation of coalitions. Any two of the voting blocs can form a majority throughout the voting and thereby achieve any outcome that they can agree on. According to our earlier analysis, with the preference order shown in Table 8.1, the amendment fails and then the unamended bill passes; in other words, the final outcome is B. Both the Republicans and the Northern Democrats prefer A to B, so it is in both their interests to band together to insure that the outcome is A rather than B. Imagine the Northern Democratic leader addressing the Republican leader: "We would like this amendment to pass. However, we know that it would, in effect, kill the bill, so we are going to vote against it. The result will then be B, your worst outcome, so you should be interested in a deal: We'll vote our beliefs on the amendment (for it) if you'll agree to vote for the whole package." Thus, the Northern Democrats, for whom "beliefs" yield their worst outcome, could rise by "calculation" to their middle outcome, and by coalition to their best.

There are two little difficulties with this coalition. First, it cannot be enforced: after the amendment passes, the Republicans can back down on their part of the deal and achieve their most favored result, though presumably such behavior will hamper any future attempts they may make to form coalitions. Another difficulty stems from coalition instability discussed in Section 8.1. Specifically, the Southern Democrats, who get their worst outcome if left out of a coalition, can approach the Republicans with a deal to achieve C, which both prefer to A. In fact, for any coalition that might form, the bloc left out can always offer one of the coalition members a deal that would improve the outcomes for both members of the proposed new coalition. In sum, although the paradox of voting is rendered inoperative by a specific voting procedure, a "paradox of coalition instability" with the same cyclical nature can arise in its place.

The order in which bills and amendments are voted on can be crucial. In the example we have been considering, the alternative A might have succeeded (even without coalitions) if it had been introduced as a separate bill. Recall that A, which we have been calling the amended bill, was a proposed law calling for federal aid to education in states not having officially segregated schools. Suppose that such a bill were introduced as a single package. Further suppose that it was understood that if A were to be defeated, a new motion would then come onto the floor simply consisting of our old B, the proposal on funds for education with nothing said about segregation. The procedure would then be that of Figure 8.5. This has been called the "successive procedure," as opposed to the "amendment procedure" given by Figure 8.4.*

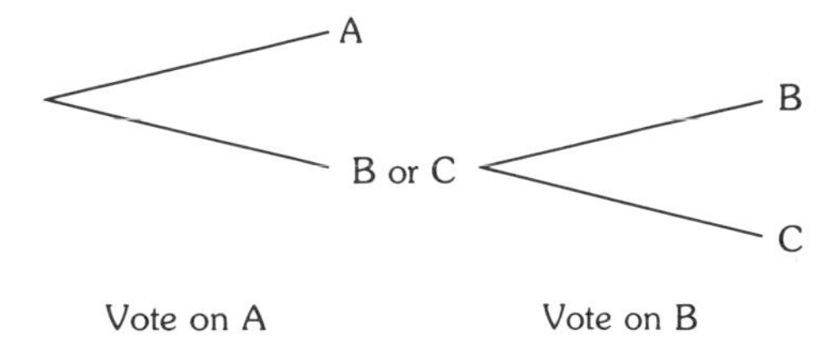

Figure 8.5 Voting procedure if A is proposed as a separate bill.

---

* Brams (1975) devotes several chapters to different kinds of voting games, and in particular he covers the material of this section in his Chapter 2.

Working backward through this procedural tree, and using the preference orders from Table 8.1, we find that on the final vote (if one occurs) B will defeat C. Therefore, at the first vote the choice is really between A, the upper branch, and B, the foreordained outcome if A is defeated. Given that the choice is, realistically, between A and B, the Northern Democrats and the Republicans will vote for A. Note that the success of A in this procedure depends crucially on the plan to introduce B if A should fail. It is also crucial that the Republicans be so informed.

Such a procedure would have been much more desirable than that of Figure 8.4, as far as the Northern Democrats are concerned. It would have enabled them to vote their beliefs with no sacrifice in the outcome. Incidentally, it would have put the Republicans on the spot in the sense that they would have had to express their true preference between A and B. Recall that with the other procedure the Republicans had a plausible rationale for supporting the amendment and then voting down the whole package. They could thus appear to be antisegregationist without any bill actually passing. With this procedure, however, if they vote for A it will pass.

One could imagine a procedural vote as to whether the proper order of voting is to be by the amendment procedure of Figure 8.4 or by the successive procedure of Figure 8.5. Assuming the preference order of Table 8.1, we can imagine that players will figure out that the amendment procedure will result in B, while the successive procedure yields A. Therefore, since two of the blocs prefer A to B, the procedural victory should go to the successive procedure with final outcome A.

Not much has been said so far about the strategic considerations of the Southern Democrats. This is because they have a dominant strategy, whichever procedure is used, and in such a case all the voting principles are in agreement. (Neither of the others has a dominant strategy with the procedure of Figure 8.4; although some of their strategies are dominated, no single strategy dominates all others, under this procedure.) Consider the preference order of the Southern Democrats, BCA, with regard to the successive procedure of Figure 8.5. At the first branch the alternatives are A versus a situation that may result in B or C. But *either* B or C is better than A, so no matter what the other players do, the Southern Democrats are better off voting against A. The key to dominance is in the phrase "no matter what the other players do." We have already seen that the decision in the final branch is strategically simple for everyone. Thus, the strategy "Vote against A, and if A loses then vote for B" is dominant for the Southern Democrats because it gives outcomes at least as good as any other strategy, no matter what the others do. Any other strategy may be regretted by the Southern Democrats.

When each branching point in the voting procedure tree splits a player's preference order so that all the outcomes that remain possible on one branch are better than all those on the other branch, then that player has a dominant strategy. The Southern Democrats' ordering is split BC/A by the first decision point in Figure 8.5. Each possibility on the bottom branch (B or C) is better than anything on the top branch (A). The amendment procedure of Figure 8.4 also splits the Southern Democrats' scale, although C straddles the split. Either alternative on their bottom branch (B or C) is at least as good as anything on the top branch (A or C), so the bottom branch is at least as good for the Southern Democrats no matter what the others may do. Their dominant strategy in this case is to vote against the amendment and then to vote for the bill just in case the amendment fails.

Although the Northern Democrats have a scale that is split (and hence a dominant strategy) in Figure 8.5, they do not have one in Figure 8.4. It is because they lack a dominant strategy in the latter case that the internal conflict can arise for them between a strategic approach and voting their beliefs. The Republicans do not have a dominant strategy in either voting procedure. In particular, the procedure of Figure 8.5 offers them a choice of A, in the middle of their preference order, versus the possibility of their best (C) or worst (B) outcome. Voting Principle 3, which refers to amendments, makes no prescription here, since A is not regarded as an amendment in the successive procedure. The Republicans could use the maximin principle to pick A, or they could vote against A on the basis of the maximax principle. However, as noted earlier, Voting Principles 1 and 2 dictate that they vote for A, since if A fails, the other two groups will combine to pass B.

Exercises

○ **1** Characteristic City has a five-member city council, and one of the five is also mayor. To pass, a bill needs a majority including the mayor or else unanimous support except for the mayor. To simplify matters it is suggested that a new system be adopted that will be equivalent to the old. It would provide that the mayor cast k votes and that a bill needs n votes to pass. For what values of k and n will the new system have the same characteristic function as the old? What is that characteristic function?

○ **2** Suppose that in Game 8.9 it is desired to form a small executive committee including all parties. Such a multi-party approach has come to be called "consociational democracy." This committee would vote by majority rule. Making the committee as small as possible, how many members should each party be allotted so that the executive committee has the same characteristic function as the legislature?

○ **3** Suppose there are three candidates, A, B, and C, and that 40 percent of the population prefers A to B to C, 15 percent prefers B to A to C, 20 percent prefers B to C to A, and 25 percent prefers C to B to A.

(a) Who would win a plurality election?

(b) If there were a run-off election, who would qualify to get into it and who would win?

(c) Show whether or not the candidate eliminated from the run-off could actually have beaten the winner of the runoff.

(d) Assuming that candidates and voters are arranged on a one-dimensional spectrum, who is the candidate of the center?

○ **4** Suppose that one of the letters A, B, C represents the status quo (no bill passes) and that the other two represent two different bills. The percentages given in exercise 3 are now taken to be groups of legislators. By what combination (if any) of procedure and strategy-selection principles can the outcome possibly be a victory for A? For B? For C? (One answer for each of A, B, and C is sufficient. Your answer may be "impossible," if that is correct.)

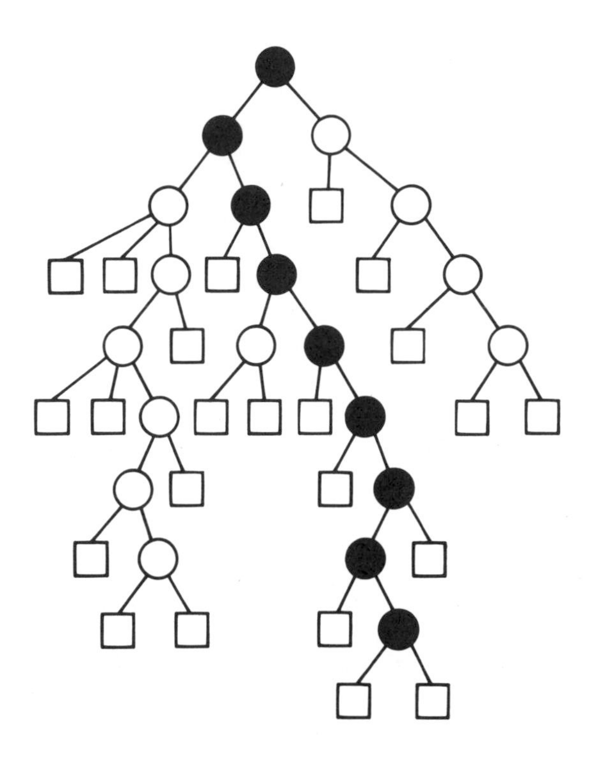

9 Experiments with Games

This chapter will highlight the most suggestive findings in the literature of experimental games. Before delving into particular experiments, however, it seems appropriate to deal with the general question, "Why the laboratory?" In other words, might not the time spent doing experiments be better spent observing naturally occurring social situations and events. The first section discusses this question and provides some perspective on the experimental approach.

The lion's share of games research has, perhaps understandably, gone to "Prisoner's Dilemma," the topic of Section 9.2. Psychological and sociological factors affecting behavior in games have turned up some provocative results, reviewed in Section 9.3. In Section 9.4 we shall turn to information, which was discussed as a "circumstance of play" in Section 5.3 and as a means of resolving dilemmas involving large groups in Section 7.3. Information turns out to have a salutary effect in a variety of games. In the last two sections we treat experimental games with more than two players; first examining the effects of group size in dilemma games, and then turning to coalition-formation in games defined by the characteristic function.

9.1 Why the Laboratory?

Although it is nice to think that game theory gives us some insight into a variety of real-life situations, a rigorous test of the theory is difficult if we cannot confirm what game is being played. To know what the game is, we need to know the participants' utilities. It is here that experimental games enter the picture, by helping to provide us with reliable knowledge of utilities. We shall return to this advantage of experiments, after first dwelling on the difficulty of determining utilities in real-life situations.

All of the various "solutions," "principles," and strategic analyses used in Chapters 3–8 are expressed in terms of players' preferences on an ordinal or interval scale. An example of how perplexing things can get when we do not know a player's true preferences is found in our analysis in the previous chapter of the Republicans' role in the Powell Amendment. Without knowing what the Republicans really preferred, we could not determine what voting principles were consistent with their actions.

In noninteractive situations it may be possible to guess a person's utilities by observing his choices If, for example, someone chooses milk rather than orange juice, then presumably milk has higher utility for him, at least at that moment. But if a union goes on strike, it is not safe to conclude that it prefers the strike to management's final offer. Bargaining power in future plays of the game is at stake here, not to mention the ability of the union leadership to stay in office. Compounding the difficulty of judging or finding out a bargainer's utilities is the fact that, as noted in Section 5.3, it is in each bargainer's interest to disguise his utilities.

To take a specific event in recent history, consider President Ford's 1975 order to seize the Cambodian vessel *Mayaguez* in order to release Americans held captive. A straightforward interpretation of Ford's utility scale in this situation would be that he preferred seizure to nonseizure. Such a preference could result from a desire to look decisive and competent. On the other hand, it is possible that the probable loss of lives involved in the seizure gave that act a low utility. The decision to order the seizure might therefore have been an attempt to gain credibility, hoping that aggressive action (with its risks) would lead to better payoffs (lower risks) in future encounters.

Experimentation under controlled laboratory conditions holds out the hope that we can overcome our ignorance of players' utility scales. If, for example, we get individuals who are not rich and have them play a game for substantial sums of money, and if we encourage them to maximize their own winnings and if we use players who are strangers to each other, then we may be justified in assuming that each player's preference among outcomes is exactly reflected by the various monetary rewards that the player gets from the outcomes. Players may still confound us by seeking equity or maximizing the difference in payoffs (see the discussion of "Spite" in Section 4.2.4), but if they do not, then we at least know what the players' utilities are and hence what game is being played.

Having thus gained confidence that we know what game is being played, we can move to an analysis of how players make strategic decisions in that game. As a result of moving to the laboratory, we not only gain a knowledge of what the payoffs are, but also the ability to control them and repeat a game of interest at will.

In addition to presenting a complex picture of utilities, real-world situations also do not repeat themselves at our command. One key to the advance of science has been the replicability of experiments: under controlled circumstances the same experiment should lead to the same results. In the real world we typically cannot cause a game to be played repeatedly under fixed conditions or with certain controlled variations that are of theoretical interest. In the laboratory, however, such things are readily arranged.

A price must be paid in moving from complex situations in the world to simple games in the laboratory. When people are put in a simple artificial situation, one might argue, they will behave in ways appropriate to a simple artificial situation, thereby revealing nothing about how they will behave in a complex real situation.*

Is this too high a price to pay? The question of artificiality has generated considerable controversy with respect to experimental games. Realism is sometimes achieved in social psychology experiments through deceit. Subjects may be misled about the aim of an experiment or even be unaware of being in an experiment at all. For example, a research assistant "accidentally" drops her papers in front of someone in a public place and observes whether the person helps. Such naturalistic circumstances presumably elicit a more reli-

* A balanced and insightful discussion of this criticism is found in Smith (1973, Chapter 1).

ably genuine response than simply showing subjects a matrix. However, since matrices give us the control we need over payoffs, the two experimental approaches are important complements to each other.

It is worth noting, moreover, that simple artificial environments have a long and successful history in many branches of science. Artificiality is a problem not just for game theory but for much of social science, since people (unlike stars and molecules) can be aware of being watched. Simplicity can be a problem if we are too careless in generalizing from the laboratory to the real world. It may be that only one aspect of a complex real situation has been subjected to experiment, or that several aspects have been studied but not their interaction, or that the real situation has been incorrectly specified. For any of these reasons it may be premature to apply certain conclusions reached in the laboratory. Nevertheless, as long as we maintain a healthy skepticism, it is not unreasonable to expect to learn something from experimentation. In fact, a number of intriguing results have been found and replicated.

9.2 Lock-in and Tit-for-Tat

Two of the most striking and suggestive results to come out of experiments with games are the "lock-in" and the effects of "tit-for-tat" strategies. Both of these effects have been extensively documented in work by Rapoport and Chammah (1965) in their studies of the repeated-play version of two-person "Prisoner's Dilemma." A lock-in consists of a long series of plays in which both players do the same thing, both or neither of them cooperating. A tit-for-tat strategy is one in which one player decides to always copy the other player's preceding choice. Since both of these effects depend crucially on repeated play, we must digress briefly to see whether it is useful to examine repeated play.

When a game is to be played more than once the structure of the overall situation is a "super-game," as noted in Section 5.2, and is therefore different from the simple game that is being repeated. Nevertheless, it is of interest to study the super-game for several reasons. For one thing, the real situations that are to be modeled allow for repeated choices among similar options. There is a new arms budget every year, labor and management renegotiate every few years, and coalition governments are periodically replaced. Another reason for repeated play is that it gives the experimental subjects an opportunity to immerse themselves in the situation. Not being game theorists, they may find it difficult to analyze all the ramifications of a matrix until they have had an opportunity to participate actively and learn what may be the consequences of various decisions. Thus, a player's first move may not tell us all there is to learn about his view of the game.

The lock-in consists of consistent CC response pairs (both players choosing C) or else consistent DD's (both players choosing D). Neither of these should surprise us. DD is the result of dominant strategies, hence plausible; CC is even better than DD for both players, hence it too is plausible. But what of the response pairs CD and DC (one player choosing C, the other D)? Since there is a rationale for C and for D, what if the two players hit on opposite rationales?

This is perfectly possible on the first play of the game. If each player flipped a weighted coin that turned up C 45 percent of the time and D 55 percent, we would expect the pairs CC and DD to turn up respectively $(.45)^2 \approx 20$ percent and $(.55)^2 \approx 30$ percent of the time, with the mixed pairs CD and DC taking up the remaining 50 percent. These figures are in fact typical of the first play. However, Rapoport and Chammah found that mixed choices drop off from 50 to 25 percent in less than 100 plays and ultimately, in 300 plays, to less than 15 percent.

Although C and D are each plausible separately, they are not plausible together since the C player with a mixed pair is getting his worst payoff and so has nothing to lose by changing. Moreover, since the D-chooser is getting maximum payoff, the C-chooser is actually encouraging "misbehavior." One might well wonder why there are any CD outcomes at all after players have had plenty of time to become familiar with the game.

CD outcomes may be preceded by CC, in which case one might imagine that one of the players deliberately exploited the other. If CD is preceded by DD then we may presume that one of the players was attempting to initiate joint cooperation. If this player persists in choosing C then perhaps he is continuing the attempt. In view of this sort of discussion, Rapoport and Chammah suggest looking at the relative frequencies of transitions, for example, $f(C|DC)$, the proportion of times a player chooses C just after having unilaterally picked D.

In view of the apparent relevance of "Prisoner's Dilemma," and since the lock-in effect suggests that the best realistic long-run objective is joint cooperation, many game theorists have wondered what one player can do to induce the other to join him in use of the dominated C choice. A policy of always picking C might seem plausible, say, to set an example. However, a policy of all C's, or for that matter one of all D's, is unresponsive to the other player's choices. If the other player cannot hope to influence you, then he will do best by always using his dominant choice. The same comments apply to any preset probabilistic mixture of C's and D's.

Only if you use a policy that is contingent on the other player's choices will it be in his interest to consider using C. The most simple and straightforward policy is the tit-for-tat strategy, which copies the other player's previous move. If you invariably give tit-for-tat, then the other player can always play C and always get his second-best payoff. This is the best he can do in the long run (provided that, as is typically the case in these experiments, the best and worst payoffs average to less than the joint-C payoff). Tit-for-tat thus seems a promising policy, provided it dawns on the other player what is happening.

The tit-for-tat, all-C, and all-D policies actually turned out, in 300-play experiments conducted by Chammah, to give results very much in line with the above comments. Chammah used a stooge player, who was told beforehand what strategy to use, as one of the subjects in each pair. The other subject was making genuine choices, unaware that his co-player was under circumstances any different from his own. Tit-for-tat elicits the most cooperation, as logically it should, with subjects using C around 75 percent of the time. It would be surprising if much cooperation were elicited by an inveterate D-chooser, and indeed little is, only about 10 percent of the time. A persistent C-choice gives intermediate results, eliciting cooperation roughly 45 percent of the time. Various probabilistic mixtures of these three policies have been tried by different experimenters (see Oskamp, 1970). In general, tit-for-tat strategies have elicited most cooperation.

A word of caution is in order. There is a certain primitive elegance to the notion of justice embodied in tit-for-tat—a C for a C, a D for a D. However, the world is not primitive, and when we emerge from the laboratory we must analyze such complex situations as whether the decision by the U.S. not to build a cruise missile is a tit-for-tat response to the decision by the Soviet Union not to build a backfire bomber, and whether the release of 10 prisoners from Cuban jails is cooperative enough to warrant ending the U.S. boycott of Cuban sugar.

Suppose two players in some situation both have a policy of tit-for-tat, and party A makes what it considers to be a cooperative move, but that move is viewed as relatively uncooperative (or not cooperative enough) by B. The stage is now set for an unending series of D choices, as each continually copies the other. Different people and different nations may have different views of what will be the likely consequences of an act, and of how harmful, beneficial, or perhaps unimportant those consequences will be. For these reasons it is always a subjective judgment how cooperative a real-world action is, and so laboratory results must be applied only with great caution.

In response to reasoning of this sort, Micko *et al.* (1977) have developed what they call "fixed-interval benevolent tit-for-tat strategies." In one of these, a stooge uses tit-for-tat unless there have been exactly four consecutive DD outcomes. Thus where tit-for-tat allows an unending string of DD's, these fixed-interval strategies can create a clear pattern: a bona fide subject who persists in D will be presented with a C-choice at every fifth playing. This provides a model of behavior that is both suggestive and easy to coordinate with. This strategy was even more effective than pure tit-for-tat, eliciting reliable cooperation earlier and from more subjects.

Just what is the experimenter studying in these games? On the first play of the game we are observing a person's reaction to a situation, and through this reaction perhaps making some tentative guesses about her decision-making processes. The situation to which she is reacting on that first play is a payoff structure over which we, as experimenter, have control. However, on the second and subsequent plays, the player is reacting not only to the game but also to the behavior of the other player. If both players are bona fide experimental subjects (not stooges who have been instructed how to choose) then their decisions become progressively more intertwined, one responding to the other. The environment of each single player is thus hopelessly out of the experimenter's control. To study individual decision-making in a systematic fashion it is thus practically essential to have one of the players be a stooge.

On the other hand, the lock-in effect, by its very nature, is antithetical to the use of a stooge. What the lock-in phenomenon shows is that despite the wide variety of decision policies that people may come up with when put into a laboratory situation, and despite the potential interactive complexity when two uninstructed people play repeated "Prisoner's Dilemma," a certain gross pattern nevertheless reliably emerges over the course of 300 plays: CD response pairs become rare for almost all subject pairs.

The lock-in effect is given the chance to emerge when, instead of studying an individual player (by making his co-player a stooge), we make the game itself the object of study. By letting many different people, of a variety of ages, occupations, sexes, personality descriptions, and nationalities all play the same game, we may find that the game fairly reliably elicits certain patterns of interaction. These patterns of interaction will depend not only on the fact that the game is, say, "Prisoner's Dilemma," but also on the precise values of the payoffs and the circumstances of play. For example, we might expect greater cooperation as the CC payoff is gradually raised, holding everything else fixed. This in fact occurs. Rapoport and Chammah used R = 1, 5, and 9 in Matrix 9.1 and obtained cooperation frequencies of .46, .63, and .73, respectively; conversely, they found

R, R	−10, 10
10, −10	−1, −1

Matrix 9.1 "Prisoner's Dilemma" (R = *R*eward for joint cooperation).

that cooperation was facilitated by raising the "price" of a DD outcome. These investigators further found that cooperation increased as T was lowered in Matrix 9.2 from 50 to 10 to 2. This result is by now expected, since lowering T is equivalent to raising (with respect to an interval scale) both the relative reward for CC and the relative "price" of DD.

1, 1	−T, T
T, −T	−1, −1

Matrix 9.2 "Prisoner's Dilemma." If players choose oppositely, the D-chooser gets T and the C-chooser gets −T.

9.3 Who You Are

What is it that American boys do more as they grow older, that Belgians do more than Americans, isolationists do more than internationalists, and power-seekers more than affiliation-seekers? Answer: they use the D strategy in "Prisoner's Dilemma." What does this tell us? This second question does not have a very clear-cut answer. For example, pairs of women use D more often than pairs of men, according to several investigators. However, Vinacke (1959) has found women to be more accommodative than men. Vinacke (1969) has attempted to reconcile these findings in an article reviewing experimental games.

> I believe, however, that the most reasonable explanation for sex differences in the [Prisoner's Dilemma] game lies in a combination of the tendency for females to be more compliant and of the demand characteristics of the situation. It is very easy for a subject to perceive a matrix game as competitive, especially under standard instructions. If females are trying hard to perform adequately, and thus to satisfy the experimenter, then they may well respond with a high incidence of competitive choices. In the broadest sense, females may actually be unusually cooperative. They may be competitive by cooperating with the experimenter!

The relationship between sex difference and cooperation in games seems, however, to be more complicated than this. In another review article, Terhune (1970) summarized sex differences in bargaining research as follows:

> 1. Women are generally less cooperative...; they tend to become involved in mutually punishing conflict deadlocks....
> 2. Women prefer straightforward accommodative solutions.... They tend to be more generous and make greater concessions....
> 3. When...exploited, women react with greater retaliation....
> 4. Men use a tit-for-tat strategy more, and tend to be more cooperative in response to a tit-for-tat strategy. Women are more cooperative if presented with cooperativeness from the beginning....
> 5. Women...often [fail] to recognize the "optimal" or "rational" strategy.

Notice that these observations refer not simply to how often people cooperate but to more subtle measures of behavior. For example, consider the observation, "When...exploited, women react with greater retaliation." This translates as, "In the play following one in which a woman has unilaterally cooperated, she is less likely to persist in cooperation than a man in similar circumstances." In symbols, the relative frequency $f(C|CD)$ is lower for women than for men. Such measures of contingent choice are examined at length by Rapoport and Chammah (1965).

People can be categorized in a variety of ways, for example, by their group (sex, nationality, age), beliefs (internationalism), or personality characteristics (need for power, affiliation, or achievement). People characterized by distrust, cynicism, and a desire to manipulate others have been studied extensively by Christie and Geis (1970), who refer to them as "Machiavellian." One might expect people with such a view of the world to be uncooperative, for cooperation requires a measure of trust. However, Tedeschi and his colleagues (1973) point out the following:

> Given the opportunity, [Machiavellian] personalities can be expected to attempt to manipulate the behavior of others for the purpose of maximizing their own rewards. Remember that a tit-for-tat strategy is the most effective one for controlling another's cooperative behavior in a mixed-motive game; then it is no surprise that... [Machiavellians] tended to play tit-for-tat....

Here again, as with sex differences, the results are conflicting and no definitive conclusions can be drawn.

Not only who you are but also who is watching can affect the way you react in a game. In an experiment conducted by Brown (1970), teenage boys played a two-part game. In the first half a stooge exploited them rather unmercifully. Each of them was told that other teenagers were watching him through a one-way mirror, and evaluating his play. Some subjects were told that the evaluators thought they were basically suckers, others that they did well under the circumstances. In the second half of the experiment, the subject had a chance to get back at the stooge. Those who were told they looked bad retaliated more than the others, even though there was a cost to themselves for doing so. The title of Brown's paper, "Face-Saving Following Experimentally Induced Embarrassment," sums up his interpretation of his findings.

9.4 What You Know

We shall consider the effects of information about payoffs in four quite different situations: "Hero," "Prisoner's Dilemma," experimental oligopolies, and a two-stage wholesaler-retailer game. Despite the structural differences among these games, a notion of cooperativeness or accommodation can be defined for each. The effect of information in each case seems to be to move the players toward equal degrees of cooperativeness.

In the wholesaler-retailer game devised by Fouraker and Siegel (1963), Wholesaler can obtain 20 items at various unit prices from \$6 to \$10. Retailer can sell them for \$11 to \$13. The most that Wholesaler can obtain and the most that Retailer can dispose of at various unit prices are shown in Table 9.1. The rules of the game are that Wholesaler sets a unit price and then Retailer decides how many items to purchase from him at that price. They are not allowed to communicate.

Table 9.1 Costs and prices in "Wholesaler-Retailer."

Wholesaler's cost		Retailer's price	
Unit price	**Quantity**	**Unit price**	**Quantity**
\$ 6	5	\$11	4
\$ 8	10	\$12	8
\$10	5	\$13	8

Part of the decision structure is shown as Tree 9.1. If Wholesaler charges a unit price of \$10.50 and all 20 items are bought, his gross income will be 20 times \$10.50, which is \$210. Subtracting his costs, $5 \times \$6 + 10 \times \$8 + 5 \times \$10 = \$160$, gives him a profit of \$50 (shown before the comma in the appropriate payoff box of the tree). The other payoffs are computed similarly.

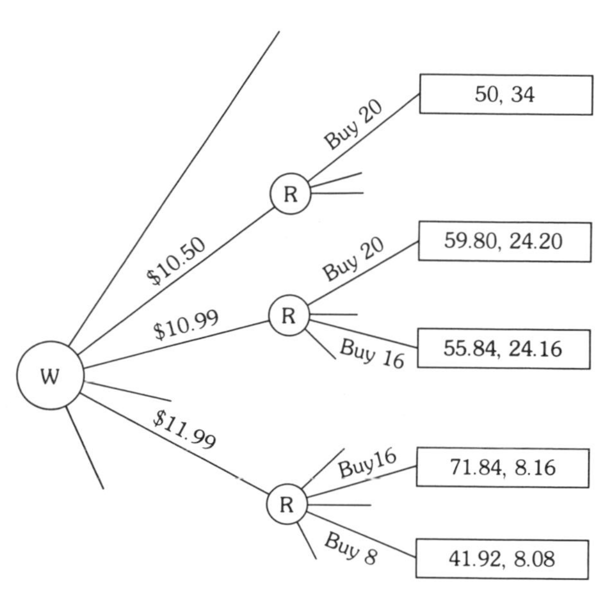

Tree 9.1 A partial tree for "Wholesaler-Retailer."

It is possible to work backward through the tree in the manner of Section 2.2, even though, as we saw in Section 4.3, this ignores threat potential in variable-sum games. Without examining the whole tree, let us note the effect of using the unit price \$11.99. At this price Wholesaler can just barely hope to entice Retailer into buying 16 items: 8 to resell at a profit of \$1.01 and 8 more to resell at a profit of just one cent (\$12 − \$11.99). If Retailer always maximizes his total profit, buying as many items as he can make a profit on, then this price of \$11.99 will give Wholesaler his best payoff, almost \$72, while Retailer will get just over \$8. This point, which is found by working backward through the tree, is called the Bowley point, and is labeled B on Graph 9.1. Its occurrence would suggest a greedy or manipulative Wholesaler and a fatalistic or compliant Retailer. We return to Retailer's decision shortly.

The dotted lines in Graph 9.1 connect sets of points that are allowed by particular prices that Wholesaler may pick. For example,

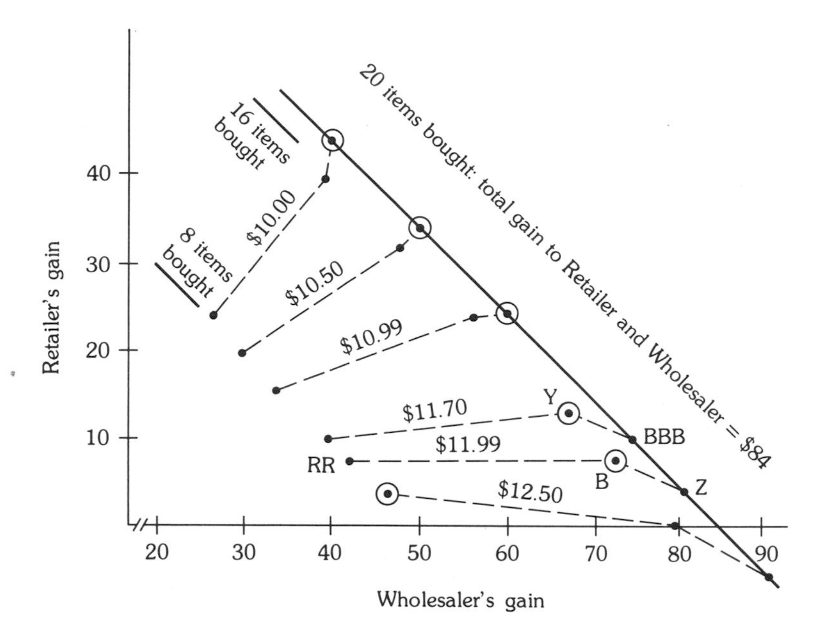

Graph 9.1 Joint-payoff graph for "Wholesaler-Retailer." The solid line is the negotiation set. Each dotted line represents the set of options to Retailer given a particular price named by Wholesaler. On each line one point is circled; this is the point Retailer will accept if he wishes to make the best of the situation Wholesaler has put him in.

the three points labeled Z, B, and RR are connected. Point Z would occur if 20 items were purchased at $11.99; point B, as noted, corresponds to 16 with that price; and RR is for 8 items at the same price. Of these, point B is the highest on the graph; that is, it is Retailer's maximum given an $11.99 price. For this reason this point is circled. Other maximum points for Retailer, at the various prices, are also circled.

The Bowley point B does not lie on the negotiation set, since point BBB is Better than B for Both (Wholesaler and Retailer). But BBB is not likely to be picked, since Retailer's maximum profit for the price $11.70 is at Y, not BBB. One might then hope that Y would constitute a solution, but a look at Graph 9.1 shows that it is not on the negotiation set.

Point B may not be so easy for Wholesaler to attain. At RR, which we may call Retailer's Revenge, the Retailer, by sacrificing 8¢ relative to the Bowley point, can cut Wholesaler's payoff by $29.92. If the game is to be played repeatedly, RR is an almost painless way for Retailer to teach Wholesaler a painful lesson.

Here at last we come back to information, the original objective of this discussion. If Wholesaler does not know Retailer's payoffs and if moreover Retailer is aware of Wholesaler's ignorance, Retailer may accede to a near-Bowley outcome, figuring that it is not really Wholesaler's fault for causing an unfair distribution of payoffs.

These last two considerations, repeated play and information, turned out to act together (though not individually) in the Fouraker–Siegel experiments to facilitate movement away from the deficient, inequitable Bowley point. Thus, with full information for both players, Retailer has less (if any) reason to put up with Wholesaler's apparent greed when, with repeated play, there is time for a lesson to be taught.

Full information also reduced exploitation in experiments on "Hero" by Guyer and Rapoport (1969). In "Hero" there are two equilibrium outcomes (see Section 4.2), each giving one player her best outcome and the other her second best. Players can do very well and gain equal outcomes in repeated plays of this symmetric game by tacitly establishing an oscillation back and forth between the two equilibria. With complete information this is exactly what happens. In the second half of a 300-play session, subjects on average use the two equilibrium outcomes about 80 percent of the time, with only about a 5 percent difference between the times spent in the two equilibria. With only partial information, the 80 percent total is again observed, but the difference between usage of one equilibrium and the other is over 40 percent. In other words, the average split was more unbalanced than 60–20.

Information effects in "Prisoner's Dilemma" were studied by Rapoport and Chammah (1965). In the "displayed matrix" condition subjects had full information about the payoff matrix and the other player's previous choice. In the "no matrix" condition subjects did not see a matrix, but were told what payoff they got at each play. The authors note that in discussions with their colleagues before the experiment:

> Some of us thought . . . that the effect of the displayed matrix would be inhibitory. We argued that in a trial and error process, the tacit solution CC would be sooner or later hit upon and would persist because of the steady positive payoffs it affords to both players in contrast to the unilateral states CD and DC, which ought to be unstable. . . and in contrast to the DD state, which is punishing. When the matrix is displayed, however, so our argument went, the dominance of the defecting strategy is a constant inhibitor against cooperating. One always is subjected to the temptation of defecting from CC to get the bigger payoff. . . . It would be better for cooperation, we thought, if these brutal facts were not explicitly before the subject's eyes.
>
> The results turned out to be exactly the opposite. . . . Instead of serving as a reminder of the prudence of choosing D,

the matrix seems to serve as a reminder that a tacit collusion is possible....

Oligopoly experiments, as reviewed by Friedman (1969), elicit very competitive behavior under conditions of partial information, but with full information a tit-for-tat pattern emerges. Here again, as with the other three examples involving very different experimental circumstances, it seems that players have a better chance of reaching some sort of accommodation if both players are fully informed of the potential consequences of their actions.

Finally, recall that in addition to knowledge of payoffs, players may receive information about one another's decisions. "To see and be seen," according to Section 7.3.1, can play a key role in promoting cooperation in real-life multi-player dilemmas. Precisely such an effect has been found by Fox and Guyer (1978) in laboratory games.

9.5 Group Size and Cooperation

To determine the effects of group size in dilemma games, one would like to use two games that are identical in every way except that one has more players than the other. Two dilemma games can be compared with respect to:

- **1** C(N), the payoff when all N players choose C.
- **2** D(O), the payoff when all choose D.
- **3** $D(n) - C(n)$, the sacrifice involved in picking C instead of D.
- **4** The benefit a player causes each other player by picking C instead of D.

Unfortunately, it is not possible to keep all of these aspects the same in two games with different numbers of players. However, "Prisoner's Dilemma" experiments have been conducted holding 1, 2, and 3 fixed. It was found that substantially greater cooperation is achieved with 3 than with 7 or 12 players (Fox and Guyer, 1977).

In an attempt to hold everything fixed but group size, another experiment used stooges to fill out the larger groups (Hamburger, 1977). Thus, one condition had two bona fide players in a two-person "Prisoner's Dilemma." In the other condition two bona fide players were joined by a stooge and they all played a three-person "Prisoner's Dilemma." The stooge (who was believed by the others to be bona fide) chose randomly with probability 1/2 of choosing C at each play of the game.

The variability caused by the stooge's behavior in the three-person game was also introduced into the two-person game by using probabilistic payoffs. That is, cooperation did not insure a particular payoff but a certain probability of getting that payoff. The three-person game was also probabilistic. The probabilities were constructed in such a way that the two bona fide players, regarded as a pair, faced exactly the same situation whichever condition they were in.

Despite the objective similarity of these two payoff structures, the players who thought they were in a three-person game cooperated less, even at the very beginning, than the players in the two-person game. Apparently, they were less hopeful that cooperative play would be effective.

Why is cooperation hard to achieve in larger groups? One way to look for an answer is to look at why cooperation does occur in two-person "Prisoner's Dilemma." There, as noted in Section 9.2, the most effective ways to elicit cooperation involve the tit-for-tat policy. But if there are, say, 11 other players and four of them choose C while seven choose D, what does it mean to give tit-for-tat? You can at most give tit-for-tat to seven others. Meanwhile, even those to whom your choice *is* a tit-for-tat will be receiving information not only about your choice but about the other players' choices as well. In the face of such potential confusion, a tit-for-tat strategy would seem unworkable.

Other explanations for lower cooperation in larger groups are possible. Suppose that some people are basically more cooperative than others. If, say, half the population fits this description, then we are much more likely to get three out of three such players for a three-person game than to get even 10 or 11 out of 12. A group consisting entirely of cooperative people might be expected to "lock in" on cooperation, as in a two-person "Prisoner's Dilemma" (Section 9.2). Whatever the reasons, the experiments show that if large groups are unable to communicate they are unlikely to use C much in "Prisoner's Dilemma" games.

9.6 Experiments on Coalition-Formation

A group of three is the smallest-size group in which some of the members can get together to benefit themselves to the exclusion of another. Thus, the smallest possible coalition game has three players, and this seems a reasonable group size to start with when conducting coalition-formation experiments; indeed, almost all the early experiments used three players.

In a typical example players are assigned "weights" of 2, 3, and 4 respectively, and any coalition with a combined weight of at least 5 can win. The winning coalition gets 1 point to divide. One may think of the weights as numbers of seats in a parliament; a coalition with a majority, in this case five seats, can form the government. Riker's theory of a minimal-winning coalition (Section 8.1.4) predicts that any two-way coalition may form, while Gamson's minimum-resources theory makes the more specific prediction that the players with weights 2 and 3 will join together. Indeed, in the earliest experiments, this pair did form most often (Vinacke and Arkoff, 1957).

It was later demonstrated, however, that if subjects become sufficiently familiar with the strategic properties of a single game, by playing that game repeatedly rather than playing a variety of games, then ultimately in most groups they realize that the "wealthy" player has no advantage and he is included as often as the other players (Kelley and Arrowood, 1960). Players in these games seem to hold beliefs that might be characterized as a self-undoing prophecy. Because the player with the highest weight is thought to deserve more, both by others and by himself, he asks more and gets avoided, thereby ultimately showing everyone that their original view was incorrect.

Four-person games allow for the possibility that a coalition may exclude someone yet still have superfluous members, something not possible with only three players. To see how often this happens, Michener and his colleagues (1975) ran a series of coalition-formation games using 24 groups of four players each. Twelve different games were played and each group played every one of them. As in the three-person games above, each player in a game was assigned a weight; a subject would have different weights in different games. They found that most coalitions were minimal, but that almost all nonminimal ones contained everyone.

For example, one of the games assigned weights of 7, 6, 5, and 2 and stipulated that a coalition needed a total weight of 12 (more than a majority) to win. Winning coalitions with no superfluous players formed in 22 of the 24 groups; the coalition 7-6 formed in 11 groups, 7-5 formed in seven groups, and 6-5-2 formed in four groups. The grand coalition of all four players formed in the remaining two groups. The possibilities 7-6-5, 7-6-2, and 7-5-2, in which there are no superfluous players, did not materialize.

Out of 288 playings (12 games, 24 groups) the grand coalition (of all players) formed 21 times, other coalitions with superfluous members formed seven times, while no coalition was formed in five games. This leaves 255 coalitions or 88 percent with no superfluous members. In contrast, Gamson's minimum-resources theory picks the 7-5 coalition in the example, thereby predicting correctly seven times out of 24 and overall 125 times out of 288 or 43 percent of the time.

The 88 and 43 percent scores for the two theories cannot be directly compared since typically there were two or three coalitions with no superfluous member but only one with minimum resources. If we predicted by random selection any one of the minimal winning coalitions in each game of this experiment, we would have been correct about 35 percent of the time. Alternatively, one could simply guess that the two largest players always unite. This actually occurs 42 percent of the time, just about as successful a prediction rate as the minimum-resources theory. The formation of coalitions of the two largest players is consistent with Gamson's rationale (Section 8.1.4) that players who do not wish to be left out in this complex situation might well try to quickly form the simplest (two players not three), most "obvious" coalition. The results thus give some support to each of the hypotheses but are not conclusive.

Exercise

○ **1** Read a few reports of research on experimental games in such journals as *Behavioral Science, Journal of Conflict Resolution, Journal of Experimental Social Psychology,* and *Journal of Personality and Social Psychology.* For guidance, it may be useful to consult Hare (1976). Pick an article that interests you and submit a photocopy of its abstract along with the following.

(a) A description of the key features of the experimental situation.

(b) A summary of the results.

(c) A discussion of why the article is of particular interest from the standpoint of theory, experimental method, or applicability to real situations.

(d) At least one suggestion for further research in the area. State whether the proposal is yours or the author's, and explain what we might hope to learn by carrying it out.

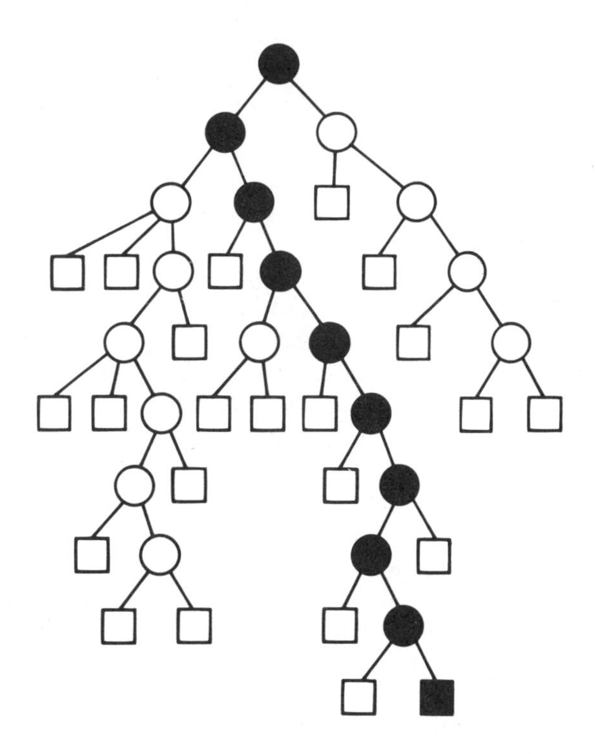

10 Cautions and Prospects

The purpose of this chapter is to put game analysis into perspective. To do this we shall first consider various points of view one might have when doing a game analysis. Next we shall take note of some pitfalls that a game analysis may fall into. The final section suggests areas of application and relevant reading in the fields of political science, economics, and psychology relevant to game theory.

10.1 Points of View

The two main endeavors of this book have been to describe interactive situations as games and to find reasonable ways those games might be played, or more briefly, analysis and solution. The first of these involves asking who the significant decision-makers are and focusing on the major decisions facing each of them. The next phase of analysis is to examine the likely consequences of various combinations of choices and to determine which among those outcomes would be preferred by the various players. In this process we have typically allowed ourselves considerable license to omit selected aspects of the situation in the name of simplicity. The end result is a model of a particular situation in the form of an interactive-decision structure, or, simply, a game.

Our other principal endeavor takes over where the first leaves off. Assuming that an appropriate game has been found, we ask what choice each player might be expected to make. Such choices are guided by the pursuit of preferred outcomes, tempered by knowledge of the possibly conflicting objectives of the other participants. Rather than try to analyze each game in isolation, we have sought to develop a set of general principles applicable to a wide class of situations. We have made so bold as to refer to some of these as "principles of rationality" and to call some of the results thereby reached "solutions." Thus, a solution means a statement of how people ought to or are likely to play, or how they would play if they followed our principles, along with a statement of the utility that would then accrue to each participant.

Just who we think we are when we go through a game analysis is important to think about. Do we envision ourselves as the grand vizier offering wise counsel to the emperor on how to win? If so, are we to assume that other players have equally wise advisors of their own? We can, for example, imagine a clever advisor to Agatha or to Bertram in Section 5.3 suggesting the various stratagems like misrepresentation and binding commitment. Or perhaps we might advise an individual in a game of "Prisoner's Dilemma" by showing him his dominant strategy. In short, we are using the term "advisor" to describe a how-to-do-it view of the game at the level of the individual player. The limitations of such a viewpoint are by now clear. We turn next to points of view that encompass all the players.

Suppose that we think of ourselves instead as detached scientists, trying to explain and predict social phenomena. If we have done well with our twin jobs of modeling and solving, that is, if we have both accurately abstracted the situation to a game and correctly analyzed what choices might reasonably or "rationally" be made in that game, then we would hope to be able to translate those choices back into real situations, where they become our prediction of what people will do. Insofar as they actually do the things we predict, we gain confidence in the general method, the particular model, and the particular strategic analysis, and we can think of ourselves as having offered an explanation of what is going on.

Glossed over in this scheme is the assumption that people are clever enough to think through the full interactive consequences of their actions, thereby finding the strategy (or complete game plan) that is "rational." Still in the role of scientist, we may examine what people actually do and find that they appear to be limited in their reasoning ability in systematic ways or that their reasoning follows other paths than those we have envisioned. Then it will be appropriate to revise or augment the theory, perhaps including psychological variables to account for variations among decision-makers.

In addition to the roles of advisor and scientist, there is that of law proposer (or political theorist). We imagine the law proposer weighing the merits of various hypothetical sets of laws (regulations, taxes, etc.) by considering what people are likely to do in specific situations, given those laws, and how desirable the end results would be. For example, in a situation with the structure of "Prisoner's Dilemma" the law proposer may suggest banning or taxing the dominant strategy.

10.2 Cautions

Game theory can yield important insights into what goes on in the world, or at least so we have claimed. But no theory or approach will yield insight if it is misapplied, so it would be well to look at some potential pitfalls. Probably the most fundamental confusion that can be made is to treat a variable-sum (mixed-motive) situation as if it were zero-sum (pure conflict). In fact, you do not even have to know game theory to avoid (or make) the mistake of blindly pursuing your own objectives without regard for the effects on others and the possibility that you could both (all) do better by coordinating your efforts.

A recent decision that bears the marks of zero-sum thinking is the building of the cruise missile in the face of its potentially destabilizing effects (see Section 4.3.3). More generally, in personal and business affairs as well as in international relations, driving too hard a bargain may mean getting no bargain or settlement at all. This is no problem for the used-car buyer, who can always go to another dealer, but it is indeed a problem for a wife and husband or for the United States and the Soviet Union, who are fairly tightly constrained to deal with each other for many purposes.

Figuring out who the players are sounds easy but may not be. For example, in the realm of international affairs a straightforward step would be to take the nations as players, as has been done in various examples in this book. Allison (1971) contends that such a selection is a common flaw in "rational actor models" that keeps us from a meaningful analysis by preventing us from examining the crucial interactions that take place *within* nations: political maneuvering, bureaucratic rivalry, etc. Such a criticism need not cause us to reject game theory, but if correct it certainly means we have chosen the players incorrectly.

It is also possible to envision a player as comprising not just one but many nations. For example, in the United States in the 1950s it was common to hear talk of "The Communists," as if communism were a monolithic international organization with centralized decision-making in Moscow. The subsequent rift between China and the Soviet Union showed that it was no longer useful—and perhaps never had been—to think of all communist nations as a single player. More recently, Egyptian President Anwar Sadat's visit to Israel, which provoked surprise and disdain in many other Arab nations, suggests that it is a mistake to regard "The Arabs" as a single player in Middle East politics.

Notice again that you do not have to be a game theorist to make mistakes of this sort. Anyone can make the mistake of failing to take into account that Egypt and, say, Saudi Arabia, have such different goals that they must be treated separately. On the other hand, there is a limit to how refined an analysis one can comprehend. "Western Europe" is often taken to be a participant in world affairs, but Italy may have different concerns than Belgium, Italian Christian Democrats differ dramatically from Italian Communists, and one could even descend to the level of party factions. At that rate, there would be hundreds of players in the game of world affairs and analysis would be hopeless. Clearly, an appropriate level of analysis must be chosen for each problem.

Turning from players to the alternatives available to them, we may note that a good mediator or a good legislative compromiser may succeed by proposing alternatives or conditions that are acceptable to the participants but that they themselves had not thought of. The game analyst, like the participant, can err by overlooking alternatives. It is also possible to confuse one alternative with another. For example, in Section 9.2 it was pointed out that the tit-for-tat policy, which is usually successful, could be disastrous if an act intended as cooperative were interpreted by the other player as being uncooperative. Such misunderstanding between players can lead to escalation of hostilities.

We would seem to be fairly well protected from confusion over players' preferences for outcomes by the existence of a nicely worked-out utility theory. One tool of utility theory, the interval scale, was discussed in Section 3.2. To make use of it we simply ask a player certain questions about his preferences among outcomes and lotteries of outcomes. Of course, if the player is a nation, a corporation, or other complex group, it is not obvious who our questions should be addressed to, unless decision-making is absolutely centralized in one person. Moreover, we have seen that it may be advantageous for a player to misrepresent his payoffs. If, despite these difficulties, we are determined to understand the world by speculating about it, we can at least estimate players' preferences on the basis of what they say and what they have done in the recent past in comparable situations. There are a number of points to keep in mind when doing this.

The notion of a diminishing marginal utility for money applies to most people. In essence, your first million dollars is worth more than your second. In the language of expected utility, you would take a sure million rather than a 50-50 shot at two million, assuming you are not rich. This can be interpreted as meaning that people have some interest in security; they will tend, for example, to buy insurance. If they are in management positions in corporations, they may be less concerned with maximizing profits than with making sure that the firm is not put in the position of having to fire people in management positions (Galbraith, 1968).

A strong assumption implicit in utility theory is that people can compare values of very different kinds of things. But in life, decisions must be made and the value of life itself enters into even the simple noninteractive decision of whether or not to use a seat belt, because there is some probability, though small, that it may save one's life. Nations, too, must deal with difficult value comparisons, involving such things as loss of life and quality of life. A government's evaluation of these things is reflected in the decisions it makes.

Utilities must incorporate players' altruism or spitefulness. We saw in Section 4.2.4 that spitefulness could turn a nonconflict game into a "Prisoner's Dilemma." On the other hand, if everyone were altruistic to the point of considering benefits to others as equal in importance to benefits to himself, then there could be no Prisoner's Dilemma.

The theory of cognitive dissonance in social psychology raises a serious difficulty for utility theory. Note that in all the games we have analyzed it has always been assumed that utilities are constant throughout a game. But this assumption would be invalid if the early moves in a tree game changed the players' preferences among outcomes. Such changes are just what is predicted by the theory of cognitive dissonance, which holds that an individual with incompatible ideas or values will try to reduce his mental conflict by rethinking one of those ideas or values. For example, a person who finds himself slipping into some undesirable outcome may try to look for its bright side (i.e., may increase his utility for it) to justify to himself the decisions he made. (For an excellent treatment of the social psychology of dissonance reduction, see Aronson, 1976, Chapter 4.)

In sum, we see that a game analysis of a situation must be carried out with caution at each stage. We must be particularly careful not to let the jargon and symbols of game theory blind us to our limitations. However, it has been noted that many of the difficulties encountered in a game analysis are intrinsic aspects of the situations, not impediments created by game theory. Indeed, it is to be hoped that game theory, by providing a conceptual framework, can give us something of a guide to those difficulties.

10.3 Prospects

All of the realistic situations used in this book to illustrate features of game theory have also represented fruitful possibilities for application of the theory. In this section I shall mention topics of current interest in political science, international relations, economics, and social psychology that are amenable to a game analysis and that the reader may profitably pursue.

Recognition of the applicability of game theory in the political arena comes from no less eminent a source than the New York State Court of Appeals. This court, in deciding whether legislative apportionment schemes satisfy the mandate of "one person, one vote," invoked a specific game-theoretic index of voting power. The court ruled that though voters may live in districts of different sizes, they must all have equal power, as measured by the index (*Ianucci* 282 N.Y.S. 2d at 507; for discussion of several voting schemes and court decisions in game-theoretic terms, see Grofman and Scarrow, 1978).

Many other types of voting schemes, in addition to representation by district, have been subjected to game analysis. These include the simple vote, weighted voting (as in shareholder meetings of corporations, where voting is weighted by the amount of stock held), and the voting scheme used in bicameral legislatures where the possibilities of veto and veto-override exist.

In Chapter 8, coalition-formation was treated without mention of such voting schemes, but the two topics interact in interesting and important ways. Coalitions may form for the purpose of trading votes. The prospects for enforcing their agreements, the combining of issues through amended bills, and other legislative phenomena are strongly affected by the rules of the legislative system, including such things as committee structure and procedural rules (mentioned in Section 8.2). These matters are well suited for game-theoretic analysis because many of the rules are precisely stated and others are fairly well agreed upon by political analysts. The reader interested in pursuing this area would do well to begin with Brams's *Game Theory and Politics* (1975).

A game-theoretic solution to a negotiation situation (Section 6.1) prescribes the final payoffs but says nothing about the process by which players come to agree on those payoffs. One such process is to seek issues of unequal importance to the conflicting parties. Then each side gets its way on an issue it considers more important in exchange for a concession on an issue it finds less important. Trading concessions in this way depends on players revealing something about their preferences, something that players may not be anxious to do in light of the stratagems described in Section 5.3 and threat potentials (Section 6.1). It is therefore of interest to consider the mediation maneuvers of Henry Kissinger, the U.S. Secretary of State, 1969–1977. A hallmark of Kissinger's technique was that he would speak to only one side at a time, jetting from capital to capital, so that his efforts were referred to as "shuttle diplomacy." Although it is hard to sort out the relative importance of Kissinger's social skills, his power a U.S. Secretary of State, and his negotiating technique, it is worth noting that shuttle diplomacy makes it possible for each side to state its priorities to the mediator without having to reveal them to the other side. From this information the mediator can formulate paired concessions (see above) that result in mutually beneficial solutions.

More recently (while the galleys for this book were being proofread), it was revealed that the September 1978 Camp David talks were conducted in this way. Specifically, after a notably unsuccessful initial three-party meeting, President Carter (the mediator) began meeting separately with Sadat of Egypt and Begin of Israel.

The economics of public choice is a natural area of study for the reader interested in the socio-economic dilemmas discussed in Chapter 7. This subdiscipline of economics applies principles of economic analysis to taxation, public spending, and governmental regulation. An excellent anthology on the subject is *Public Expenditure and Policy Analysis* by Haveman and Margolis (1976). Dilemmas within large groups must be resolved by the group as a whole, typically by governmental action. The geopolitical distribution of the players determines the appropriate level of government, e.g., city, state, or national (on this point see Tullock's very readable book, *Private Wants, Public Means,* 1970). Of course, even after an analysis of a situation suggests what ought to be done, there is still the question of how the decision-making processes of a particular political system will operate. Therefore, the dilemmas in Chapter 7, including the resolution techniques of Section 7.3, should be studied with a view to negotiation (Section 6.1) and political processes (Chapter 8).

Social psychology includes the study of prosocial behavior, defined as any act that aids another person (or persons) without any tangible benefit to the actor, for example, helping an injured stranger. Since this definition also fits the C-choice in "Prisoner's Dilemma," prosocial behavior provides an important perspective on the behavior of individuals in dilemma situations. A good introduction to experiments on prosocial behavior is Chapter 8 of *Social Psychology* by Baron and Byrne (1977). Hare's (1976) prodigious handbook provides a coherent framework for small-group research, thereby showing how, in particular, experimental games fit into the broad aims of social psychology.

My final suggestion for reading is a good newspaper. If this book has achieved its purpose, the reader will find that game analysis provides insight, or at least interesting speculation, on what is going on in the world.

Bibliography

Allison, G. T. 1971. *Essence of Decision: Explaining the Cuban Missile Crisis.* Boston: Little, Brown.

Aronson, E. 1976. *The Social Animal.* San Francisco: W. H. Freeman and Company.

Aumann, R. J., and Maschler, M. 1964. The bargaining set for cooperative games. In *Advances in Game Theory* (Annals of Mathematics Study 52), M. Dresher, L. S. Shapley, and A. W. Tucker, eds. Princeton, NJ: Princeton University Press.

Baron, R. A., and Byrne, D. 1977. *Social Psychology.* Boston: Allyn and Bacon.

Berne, E. 1964. *Games People Play.* New York: Grove.

Brams, S. J. 1975. *Game Theory and Politics.* New York: Free Press.

Brown, B. R. 1970. Face-saving following experimentally induced embarrassment. *J. Experimental Social Psychology, 6:*225–271.

Caplow, T. 1956. A theory of coalitions in the triad. *American Sociological Review, 21:*489–493.

Christie, R., and Geis, F. L. 1970. *Studies in Machiavellianism.* New York: Academic Press.

Davis, O. A., and Whinston, A. B. 1961. The economics of urban renewal. *Law and Contemporary Problems, 26:*105–117.

De Swaan, A. 1973. *Coalition Theories and Cabinet Formation.* New York: Elsevier.

Dorfman, R., and Dorfman, N. S. 1972. *Economics of the Environment.* New York: Norton.

Downs, A. 1957. *An Economic Theory of Democracy.* New York: Harper & Row.

Edelman, J. M. 1964. *The Symbolic Uses of Politics.* Urbana: University of Illinois Press.

Fouraker, L. E., and Siegel, S. 1963. *Bargaining Behavior.* New York: McGraw-Hill.

Fox, J., and Guyer, M. 1977. Group size and others' strategy in an N-person game. *J. Conflict Resolution, 21*(2):323.

Fox, J., and Guyer, M. In press. "Public" choice and cooperation in N-person prisoner's dilemma. *J. Conflict Resolution, 22.*

Friedman, J. W. 1969. On experimental research in oligopoly. *The Review of Economics, 36:*399–415.

Galbraith, J. K. 1968. *The New Industrial State.* London: Hamish Hamilton.

Gamson, W. A. 1961. A theory of coalition formation. *American Sociological Review, 26:*373–382.

———. 1972. SIMSOC: Establishing social order in a simulated society. In *Simulation and Gaming in Social Science,* M. Inbar and C. S. Stoll, eds. New York: Free Press.

Grether, D. M., and Plott, C. R. In press. Economic theory of choice and the preference reversal phenomenon. *American Economic Review.*

Groennings, S., Kelley, E. W., and Lieserson, M., eds. 1970. *The Study of Coalition Behavior.* New York: Holt, Rinehart and Winston.

Grofman, B., and Scarrow, H. 1978. Game theory and the U.S. courts. Chapter V presented at the International Conference on Applied Game Theory, Institute for Advanced Studies, Vienna, June 13–16, 1978.

Guyer, M. J., and Rapoport, A. 1969. Information effects in two mixed-motive games. *Behavioral Science, 14:*467–482.

Hamburger, H. 1973a. N-person prisoner's dilemma. *J. Mathematical Sociology, 3:*27–48.

———. 1973b. Dilemmas and sure things. (Review of *Paradoxes of Rationality* by N. Howard.) *Science, 180:*195–196.

———. 1977. Dynamics of cooperation in take-some games. In *Mathematical Models for Social Psychology,* W. H. Kempf and B. H. Repp, eds. Bern: Hans Huber.

Hare, A. P. 1976. *Handbook of Small Group Research,* 2nd ed. New York: Free Press.

Haveman, R. H., and Margolis, J. 1977. *Public Expenditure and Policy Analysis.* Chicago: Rand McNally.

Howard, N. 1971. *Paradoxes of Rationality: Theory of Metagames and Political Behavior.* Cambridge: MIT Press.

Inbar, M., and Stoll, C. S., eds. 1972. *Simulation and Gaming in Social Science.* New York: Free Press.

Kahan, J. P. 1974. Rationality, the prisoner's dilemma, and population. *J. Social Issues, 30*(4):189–209.

Kahnemann, D., and Tversky, A. In press. Prospect theory: An analysis of decision under risk. *Econometrica*.

Kelley, H. H., and Arrowood, A. J. 1960. Coalitions in the triad: Critique and experiment. *Sociometry, 23:*231–244.

Latane, B., and Darley, J. M. 1969. Bystander "apathy." *American Scientist, 57:*244–268.

Luce, R. D., and Raiffa, H. 1957. *Games and Decisions.* New York: Wiley.

Mann, L. 1973. Learning to live with lines. In *Urbanman: The Psychology of Urban Survival,* J. Helmer and N. A. Eddington, eds. New York: Free Press.

Michener, H. A., Fleishman, J. A., Vaske, J. J., and Statza, G. R. 1975. Minimum resource and pivotal power theories: A competitive test in 4-person coalition situations. *J. Conflict Resolution, 19:*89–107.

Micko, H.-C., Brückner, G., and Ratzke, H. 1977. Theories and strategies for prisoner's dilemma. In *Mathematical Models for Social Psychology,* W. H. Kempf and B. H. Repp, eds. Bern: Hans Huber.

Nash, J. F. 1950. The bargaining problem. *Econometrica, 18:*155–162.

Oskamp, S. 1970. Comparison of strategy effects in the prisoner's dilemma and other mixed-motive games. *Proceedings,* 78th Annual Convention, American Psychological Assn., *5:*433–434.

Rapoport, A. 1966. *Two-Person Game Theory.* Ann Arbor: University of Michigan Press.

Rapoport, A., and Chammah, A. M. 1965. *Prisoner's Dilemma.* Ann Arbor: University of Michigan Press.

Rapoport, A., and Guyer, M. J. 1966. A taxonomy of 2 × 2 games. *General Systems, 11:*203–214.

Richardson, H. W. 1971. *Urban Economics.* Harmondsworth, Middlesex, England: Penguin.

Riker, W. H. 1962. *The Theory of Political Coalitions.* New Haven: Yale University Press.

——— 1965. Arrow's theorem and some examples of the paradox of voting. In *Mathematical Applications in Political Science,* J. M. Claunch, ed. Dallas: Arnold Foundation of Southern Methodist University.

Schelling, T. C. 1963. *The Strategy of Conflict.* Cambridge: Harvard University Press.

———. 1973a. On the ecology of micromotives. *The Public Interest, 25:*61–98.

———. 1973b. Hockey Helmets, concealed weapons, and daylight saving. *J. Conflict Resolution, 17:*381–428.

Shubik, M. 1975. *Games for Society, Business and War: Towards a Theory of Gaming.* New York: Elsevier.

Singleton, R. R., and Tyndall, W. F. 1974. *Games and Programs: Mathematics for Modeling.* San Francisco: W. H. Freeman and Company.

Smith, P. B. 1973. *Groups within Organizations.* London: Harper & Row.

Smolensky, E., Becker, S., and Molotch, H. 1968. The prisoner's dilemma and ghetto expansion. *Land Economics, 44*(4):419–430.

Tedeschi, J. T., Schlenker, B. R., and Bonona, T. V. 1973. *Conflict, Power and Games: The Experimental Study of Interpersonal Relations.* Chicago: Aldine.

Terhune, K. W. 1968. Motives, situation, and interpersonal conflict within prisoner's dilemma. *J. Personality and Social Psychology Monograph Supplement, 8*(3, part 2).

Thompson, D. N. 1973. *The Economics of Environmental Protection.* Cambridge: Winthrop.

Tullock, G. 1970. *Private Wants, Public Means.* New York: Basic Books.

Vickery, W. 1959. Self-policing policies of certain imputation sets. In *Contributions to the Theory of Games IV,* A. W. Tucker and R. D. Luce, eds. Princeton, NJ: Princeton University Press.

Vinacke, W. E. 1959. Sex roles in a three-person game. *Sociometry, 22:*343–360.

———. 1969. Variables in experimental games: Toward a field theory. *Psychological Bulletin, 71:*293–318.

Vinacke, W. E., and Arkoff, A. 1957. Experimental study of coalitions in the tirad. *American Sociological Review, 22:*406–415.

Von Neumann, J., and Morgenstern, O. 1944. *Theory of Games and Economic Behavior.* Princeton, NJ: Princeton University Press.

Wedgwood, C. V. 1966. *A Coffin for King Charles.* New York: Time Life.

Wolfe, T. 1970. *Radical Chic and Mau-Mauing the Flak-Catchers.* New York: Farrar, Strauss, and Giroux.

Index